T0318731

Biology of Life

Biology of Life
Biochemistry, Physiology and Philosophy

Laurence A. Cole
Director, USA hCG Reference Service, Santa Fe, New Mexico, USA

AMSTERDAM • BOSTON • HEIDELBERG • LONDON
NEW YORK • OXFORD • PARIS • SAN DIEGO
SAN FRANCISCO • SINGAPORE • SYDNEY • TOKYO

Academic Press is an imprint of Elsevier

Academic Press is an imprint of Elsevier
125 London Wall, London EC2Y 5AS, United Kingdom
525 B Street, Suite 1800, San Diego, CA 92101-4495, United States
50 Hampshire Street, 5th Floor, Cambridge, MA 02139, United States
The Boulevard, Langford Lane, Kidlington, Oxford OX5 1GB, United Kingdom

Notices
Knowledge and best practice in this field are constantly changing. As new research and
experience broaden our understanding, changes in research methods, professional practices,
or medical treatment may become necessary.

Practitioners and researchers must always rely on their own experience and knowledge in
evaluating and using any information, methods, compounds, or experiments described herein.
In using such information or methods they should be mindful of their own safety and the safety
of others, including parties for whom they have a professional responsibility.

To the fullest extent of the law, neither the Publisher nor the authors, contributors, or editors,
assume any liability for any injury and/or damage to persons or property as a matter of products
liability, negligence or otherwise, or from any use or operation of any methods, products, instructions,
or ideas contained in the material herein.

Library of Congress Cataloging-in-Publication Data
A catalog record for this book is available from the Library of Congress

British Library Cataloguing-in-Publication Data
A catalogue record for this book is available from the British Library

ISBN: 978-0-12-809685-7

For information on all Academic Press publications
visit our website at https://www.elsevier.com/

  Working together
to grow libraries in
developing countries

www.elsevier.com • www.bookaid.org

Publisher: Sara Tenney
Acquisition Editor: Sara Tenney
Editorial Project Manager: Pat Gonzalez
Production Project Manager: Julia Haynes
Designer: Mark Rogers

Typeset by TNQ Books and Journals

Contents

Introduction   ix

Section I
How Life Started   **1**

1. **The Physics and Chemistry of Planet Earth**   **3**

    The Expanding Universe   3
    Birth of the Solar System   4
    Birth of Planet Earth   6
    The Geology and Chemistry of Planet Earth   8
    References   10

2. **How Life Started**   **11**

    References   17

3. **RNA, Ribozymes, and Early Life Forms**   **19**

    References   22

4. **Amino Acids and Peptides, Essential Ingredients of Life** **25**

    References   27

5. **Biochemistry of Early Life**   **29**

    References   34

6. **Evolutionary History of Planet Earth**   **37**

    Oxygen Atmosphere   37
    Prokaryotic and Eukaryotic Cells   37
    Multicellular Organisms   42
    References   42

7. **What Is Life?**   **45**

    References   47

Section II
DNA and RNA   49

8.   The Evolution of DNA and the Genetic Code   51

Evolution of DNA   51
Evolution of the Genetic Code   52
References   54
Further Reading   54

9.   DNA Biology: DNA Replication, Transcription, and Translation   55

Replication   58
Transcription   59
mRNA Splicing   60
Translation of mRNA   61
Genes and Chromosomes   62
Further Reading   62

Section III
Energetics   63

10. Adenosine Triphosphate Energetics   65

Oxidative Phosphorylation   66
Feeding Pathways to Oxidative Phosphorylation   68
Use of Adenosine Triphosphate Energy   71
Adenosine Triphosphate and the Heart   72
Photosynthesis and the Generation of Oxygen and Adenosine
    Triphosphate   74
Cytoplasmic Membrane Electron Transport and Adenosine
    Triphosphate Synthesis in Bacteria   75
Evolution of Energetics   76
Energetics and Life   77
References   77

11. Adenosine Triphosphate Synthase   79

Further Reading   82

Section IV
4.5 Billion Years Evolution   83

12. Evolution Timeline   85

Further Reading   92

13. Evolution of Chemical, Prokaryotic, and Eukaryotic Life 93

Chemical Life   93
Prokaryotic Life   93

Eukaryote Evolution From Prokaryotes   96
Eukaryotic Life   97
Further Reading   99

14.   Animal Evolution, Small Brain and Advanced
      Brain   **101**

      Animal and Mammal Evolution   101
      Small Brain Animals   101
      Advanced Brain Animals   103
      Further Reading   104

15.   The Evolution of Humans   **105**

      The Human Brain   107
      Bipedalism   113
      References   115

16.   Human Development   **119**

      Language   119
      Human Pseudoevolution   123
      References   124

Section V
Human Development   **125**

17.   Human Female Oogenesis   **127**

      Oogenesis   127
      Ovulation   130
      Luteogenesis   131
      References   133
      Further Reading   133

18.   Human Male Spermatogenesis   **135**

      Mitotic Proliferation   135
      Meiotic Division   135
      Cytodifferentiation and Packaging of Sperm Cells 136
      Endocrine Control of Spermatogenesis   139
      Maturation of Spermatozoa   139
      Seminal Fluids   140
      Further Reading   141

19.   Sperm Activation, Fertilization, Morula, Blastocyst
      Formation, and Twinning   **143**

      Sperm Activation   143
      Spermatozoa Capacitation   143

Spermatozoa Acrosome Reaction   145
Ovum Propulsion   145
Fertilization   146
Morula and Blastocyst Formation   147
Twinning   148
Further Reading   150

20. **Multiple Human Chorionic Gonadotropin
     Molecules**   **151**

Five or Six Completely Independent hCG
     Molecules   151
References   155

21. **Implantation and Pregnancy Failure**   **157**

Implantation   157
Pregnancy Failure   158
References   161

22. **Hemochorial Placentation**   **163**

References   166

23. **Human Life, Development of the Human Brain** **167**

References   171
Further Reading   171

24. **Human Life**   **173**

Index   177

Introduction

My name is Larry Cole or Laurence A. Cole PhD to be formal, and I introduce you to Biology of Life: Biochemistry, Physiology and Philosophy. I have had an unusual life plagued by science and illness. When I was 21 years old, in 1974, I was in my third year of medical school at Southampton University in England. I am not sure that I really fitted medical school in that I found life in medicine too regimented, not scientific, or not calling on the brain to solve people's medical problems. One solved medical problems only by established or agreed upon methods.

At that point, I had a nasty pain in the face and saw a dentist who could not attribute my problems to any infected tooth. The next day I saw a neurologist that the dentist recommended who could not find a neurological problem in my face and sent me straight back to another dentist. The extreme pain in the face continued. Once again I was told by the second dentist that the pain was neurological and nothing to do with my teeth. On the fourth day, frustrated, I saw another neurologist. The neurologist considered trigeminal neuralgia but wanted me to see the dentist that he recommended. The pain went on and on and was never treated. It was in this dentist's office waiting room that I had a massive grand mal seizure and entered into a deep coma.

I was told that I was taken to the nearby emergency room at Ealing Memorial Hospital with a high fever and in a coma. They diagnosed meningitis based around the fever and loaded me up with intravenous antibiotics. When 4 days later I still had a high fever, 105°F, and was still in a coma, they considered my case hopeless and transferred me to Atkinson Morley neurosurgical hospital in Wimbledon, London. I am told that I arose about two months later at Atkinson Morley Hospital. Two months later I still had a fever and doctors placed me in an iced bath. I arose from the coma shortly after the ice bath. I spent almost one year in hospital, fully recovered but having severe brain damage. My right temporal lobe and right hippocampus were totally destroyed or scarred beyond recognition and my memory as a left-handed person using the right lobes of the brain was almost zero.

I was told by Donald Acheson MD, dean of the medical school that I was in no fit shape to finish medical school or to be a physician. I felt very lost. I met my wife Linda in London and went with her to live in Milwaukee, WI USA. In Milwaukee, after working for a year as a clerk in a window factory, I was determined to return to science. I enrolled in the Medical College at Marquette

University, Wisconsin to study for a PhD in Biochemistry. What happened to me next is an amazing story. In the first year I had to complete a one year General Biochemistry full time course.

With my destroyed right side of the brain memory, everything went in one ear and straight out the other ear. I flunked every test and failed every exam. I am extremely lucky that the Biochemistry Chairman, Alan Mehler PhD had a lot of faith in me and did not throw me out. I told him everything that happened to me and he allowed me to retake the flunked year.

I restarted the year scared that I was wasting my time and that I was just going to flunk everything again. Then something very strange happened. I came home from a day of taking notes in biochemistry classes. I had supper with Linda and then went into the dining room and started to read over my day's notes. For some reason, maybe I was enticed by God, I started to shout my notes out aloud, and kept shouting them for most of the evening. Something very strange was happening. I actually started memorizing the words that I was shouting. I kept doing this every evening after I came home. In October 1978, I faced the first exam. Somehow I made it and I passed it. To cut a long story short, I passed the one year course with flying colors. I passed my comprehensive PhD exam with flying colors and completed my research with flying colors. In fact, after just 4 years I was awarded a PhD with flying colors.

I then went to the University of Michigan in Ann Arbor to complete a postdoctoral research fellowship. Somehow, I applied to the National Institutes of Health and the American Cancer Society for three very large research grants. I was awarded all three, million dollar, five year grants. Having three research grants I was snatched up by Yale University and became an Assistant Professor and the Associate Professor of Obstetrics and Gynecology, and Biochemistry.

How did I go from having a learning disability, to earning a PhD, to being hired as a professor by one of the top universities in the country? I went to see a psychiatrist and a neurosurgeon to try and discover what had happened to me. Yes, I had destroyed in England the key memory system of the right side of my brain, the active or principal side of my brain. At the Medical College of Wisconsin I somehow, by means never heard of, activated the left side of my brain, a one in a million possibility as the physicians related to me, and then obtained a PhD and a Yale University professorship with flying colors.

This is my life story. If you are interested in reading about it further please buy a book called "Larry's Left Side," ISBN 978-1-934675-49-6. It talks about me going to medical school, my stroke and coma incident, and about how I reactivated my brain while working on my PhD. It is now 2016 and I look back on my life. Still the right side of my brain which I use for general purpose is useless. I can watch a movie 10 times over and still not remember a scene. The left side of my brain is somehow reserved just for science. It still remembers all those biochemical pathways that I shouted to myself in evenings 37 years ago. The left side of my brain debates science 16 h every day. It is socially boring but never stops thinking. The left side of my brain is me, the scientist.

The origins of life and the origins of advanced human life are subjects that I think and theorize about a lot. It is my pleasure to use that left side of my brain to present my book "Biology of Life" here. I consider how life came about, how life evolved, and how human or advanced life came about. I make two clear conclusions in this book and propose multiple theories and hypotheses. One conclusion is posed by the "Chapter 7, What Is Life." This chapter looks back at previous chapters and addresses the issue of how can we identify life. If a NASA spacecraft lands on a distant planet how can we truly identify if life-forms exist. The second conclusion is posed by the "Chapter 24, Human Life." This addresses the issue of what signifies human life. Before a state considers that human fertilization constitutes human life, and destruction of a fertilized being is abortion or murder they need to read this unique chapter.

I am a keen photographer and present some of my abstract and international child photographs as the cover and as section heads in this book. This book is very much also a product of Larry's left side. Larry's left side that cannot stop thinking about science or about the wonders of the world.

With my history of amazing medical events I cannot stop thinking about and thanking God. While this book is about science and incredible science events, I can only attribute all the science to God. While God is not discussed in this book, God is originator of all pure science, it all has to be considered as God's plans.

This book examines what is life? How life first started (Section I); it looks carefully at RNA and DNA in the development of life-forms (Section II); at the importance of life energetics in all life-forms (Section III); it then carefully examines evolution and how humans developed from chemical life-forms (Section IV); and finally, the book examines human life, how human life emerges and how a human baby comes about (Section V).

Section I

How Life Started

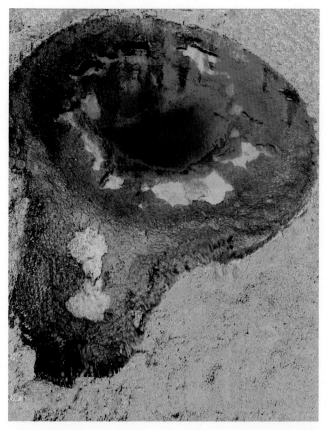

Picture taken by Laurence Cole. Geyser at Yellowstone National Park, 2007.

Chapter 1

The Physics and Chemistry of Plant Earth

Before we can fairly look at how life started. We must examine the physics and astronomical history of how the earth came about, and the chemistry of what the Earth was like when life first came about.

THE EXPANDING UNIVERSE

The universe first came about a very long time ago, with the Big Bang. I am told that according to astronomers and physicists all over the world, the generally shared theory was the expanding universe concept. The name, the Big Bang Theory came about in 1950. This was first conceived in a BBC radio show featuring Sir Fred Hoyle. I use the name Big Bang Theory but science proves that no big bang, noise, or explosion was involved in the formation of the universe. Why was it called the Big Bang Theory when no big bang ever occurred?

The Big Bang Theory considers that the universe is continuously expanding at a set mathematical rate. This has been very much confirmed in recent years by observations made by the Hubble telescope in space. The universe is clearly expanding and stars are receding or moving slowly further and further apart. The Big Bang Theory as it is now known is also confirmed by Einstein's relativity and gravity theories. In 1964, Penzias and Wilson serendipitously discovered that cosmic background radiation [1], a signal in the microwave band, is actually energy left over by the Big Bang Theory. Their discovery provided substantial confirmation of the Big Bang model [1]. The radiation was found to be consistent with the presence of dark energy or energy of the expanding universe present in all directions.

Independent lines of evidence from supernovae and the Penzias and Wilson cosmic background radiation imply that the universe today is dominated by a mysterious form of energy known as dark energy, which apparently permeates all of space. When the universe was very young, it was likely infused with dark energy, but with less space and everything closer together, so that gravity predominated, and it was slowly braking the expansion of the universe. Eventually, after numerous billion years of expansion, the growing abundance of dark energy caused the expansion of the universe.

Biology of Life. http://dx.doi.org/10.1016/B978-0-12-809685-7.00001-0

From the Big Bang Theory and estimates of the mathematical rate of expansion of the universe, the age of the universe has been estimated. It is predicted that the universe was at first nothing or completely nonexistent, and then 13.798 billion years ago (BYA) it suddenly became something, an infinitely hot (billions of degrees centigrade) and infinitely concentrated (billions of kilograms per liter) form. The form was so hot and so concentrated that neither atoms nor particles existed in it. After the initial expansion and cooling, the mass permitted these subatomic particles to form. Slowly, after an estimated 379,000 years of cooling and expansion, electrons, protons, and neutrons formed atoms (mostly hydrogen). The Higgs boson particle gave these particles a mass or made solids. Giant clouds formed from primordial elements which later coalesced through gravity to form stars and galaxies. This expansion is still ongoing 13.798 billion years after everything started.

The big question that physics has never answered is, what was the nothing that existed before the Big Bang? In general theists attribute the nothing becoming something to an intelligent God. Atheists attribute the origin to a natural event. Mathematically, $0 = +1 + -1$ or nothing equals a matching concentration of (+) energy and (−) energy. Considering this mathematics in the beginning was nothing, and after the Big Bang was a mixture of plus energy and minus energy, or mathematically nothing. Atoms were later made as an equal distribution of plus energy or protons and minus energy or electrons and neutral neutrons, and the universe exists today as a mixture of these particles. Under this hypothesis there was nothing (0) in the beginning and there is still nothing (+1 and −1) today. This may be the simplest explanation of how the expanding universe came about from nothing.

Dark energy is an unknown form of energy which is hypothesized to permeate all of space, which resulted from the Big Bang. Dark energy permeates everywhere in the universe fueling the ongoing expansion of the universe using Einstein's theory connecting energy and mass (energy = mass × speed of light squared, $e = mc^2$). According to the Planck mission team, and based on the standard model of cosmology, on a mass–energy equivalence basis, the observable universe contains 95.1% dark matter and dark energy, and just 4.9% ordinary matter and ordinary energy. Fig. 1.1 illustrates the concept of dark energy and the expanding universe.

BIRTH OF THE SOLAR SYSTEM

Under the Solar Nebular Disk Model (SNDM), stars were formed from massive dense clouds of molecular hydrogen formed by the expansion of the universe [2,3]. These clouds were gravitationally unstable structures in which hydrogen matter coalesced into dense hydrogen clumps [2,3]. The Solar system was formed from the gravitational collapse of giant dense molecular clouds of hydrogen [2,3].

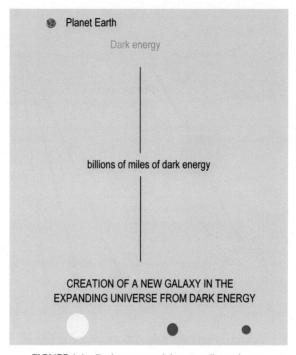

FIGURE 1.1   Dark energy and the expanding universe.

The explosion of a supernova or of a burnt-out star near to the site of the Solar system triggered the break of a cloud of molecular hydrogen. The waves of energy from the supernova explosion cleaved and squeezed the molecular cloud of hydrogen. This initiated a reaction in which the cloud became hotter and denser. The center of the cloud became so hot that it broke up and formed a protostar or the primitive Sun. The primitive Sun or protostar continued to grow by accretion of gas and dust from the molecular cloud. The side pieces of the molecular cloud cooled down forming massive hydrogen pieces giving birth to outer cold gas planets, Jupiter, Saturn, Uranus, and Neptune, composed primarily of super cold hydrogen.

In the massive protostar, the Sun, the core temperature reached approximately 10 million degrees kelvin. This initiated a proton–proton chain reaction allowing hydrogen to fuse, first to deuterium and then to helium or the onset of nuclear fusion. The nuclear fusion is the power, heat, or energy of the sun. This is how the Solar system was formed with the Sun as the chief gravitational force at the center of the galaxy. If a less massive protostar had been formed brown dwarfs would has resulted in which nuclear fusion failed to occur.

Multiple natural sciences suggest that the Sun and Solar system were formed 4.6 BYA . Fig. 1.2 illustrates the initial Solar system formation. The Milky Way,

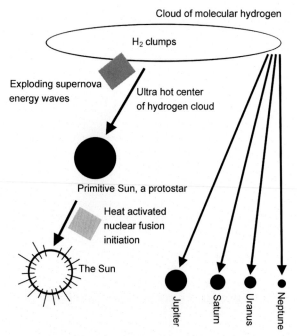

FIGURE 1.2 The Solar System formation [2,3].

our galaxy, is a disk shaped structure containing 100–400 billion stars like the sun. There are many galaxies in the universe. Planet Earth was seemingly constructed immediately after the Solar system was formed, being born about 60 million years after the construction of the Solar system. Radiation half-life dates planet Earth to 4.54 BYA.

BIRTH OF PLANET EARTH

According to the SNDM, rocky planets like Earth, Mars, Venus, and Mercury form in the inner part of the protoplanetary disk, where temperatures are higher due to the nearby Sun.

Earth began from the same molecular hydrogen cloud that formed the sun. The cloud contained dust as well as hydrogen [2,3]. After the Solar system was formed, the dust grains in the remaining broken-up molecular clouds started to stick to each other forming planetesimals, bodies roughly 1 km in width. Coagulation continued as mass accumulated, and gravity started attracting masses together. Over about 10,000 years, the bodies grew to approximately 100–1000 km in diameter or became an asteroid. Asteroids exert gravity, and occasionally these bodies impact each other to form larger bodies or embryo planets about 500–5000 km in diameter. Asteroids and embryo planets which merged together over 50 million years, formed planets

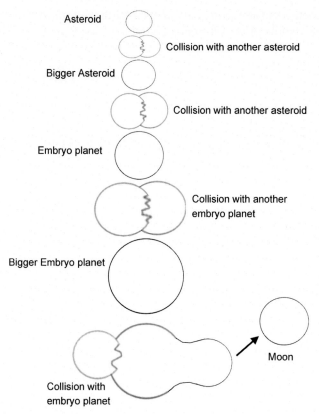

Asteroid

Collision with another asteroid

Bigger Asteroid

Collision with another asteroid

Embryo planet

Collision with another
embryo planet

Bigger Embryo planet

Moon

Collision with
embryo planet

FIGURE 1.3 Asteroid merging with other asteroids, embryo planets merging with other embryo planets, and the formation of the Moon.

about the size of Earth (Fig. 1.3). The planet Earth is 12,713 km in diameter or 7926 miles in diameter. Some of the asteroids and planet embryos are thought to have brought water to Earth since water has been demonstrated on the earliest planet Earth.

It is thought that the Moon came about from one of the later planet embryo collisions with the growing planet Earth [4,5]. This hypothesis requires a collision between a body about 10% the present size of the Earth and planet Earth. The energy involved in this collision is impressive; trillions of tons of material which would have been vaporized and melted. In parts of the Earth, the temperature would have risen to 10,000°C. Then, in response to this massive collision, the liquid Earth would have released at the opposite side of the collision a liquid body about the size of the Moon [4,5]; like in a Newton's cradle, the ball hits the Earth and the Earth releases a ball (Fig. 1.3). Multiple studies of the moon show that the Moon and planet Earth share common cores and likely common origins.

Radioactive half-life dating shows that the Earth was formed about 4.54 BYA. Over time, the planet cooled from a hot liquid mass at around 10,000°C and formed a solid crust, allowing liquid water to exist on the surface. The first life forms appeared on earth between 3.8 and 3.5 BYA. The bigger question may be why is Earth the only planet in the Solar system with abundant water? All rocky planets in the Solar system, Earth, Mars, Venus, and Mercury contain a similar composition yet only Earth supports life. Could it be that since Earth, when under construction, was invaded by asteroids and planet embryos containing water, and has abundant water, that Earth is the only planet that can support life.

THE GEOLOGY AND CHEMISTRY OF PLANET EARTH

Earth formed 4.54 BYA. The age of the earth is approximately one-third of the age of the universe. Earth formed as a hot molten mass from embryo planet collisions which led to extreme volcanic conditions. Over approximately 300 million years, the Earth cooled down and formed a sold crust. Geological changes have been constantly occurring on Earth. The process of plate tectonics has played a major role in shaping the Earth, its oceans and continents. The biosphere in turn has had a significant affect on Earth's atmosphere and other abiotic conditions such as the formation of the ozone layer, the proliferation of oxygen, and the creation of soil.

The first eon in Earth's history was called the Hadean eon. The oldest rocks on Earth date to 4 BYA and the oldest detrital zircon crystals in rocks date to 4.4 BYA, soon after formation of the Earth's cooled crust [6]. The zircon crystals prove that water was on Earth around the time of its formation [6]. The Giant Impact Hypothesis suggests that at this time, 4.4 BYA, the early Earth was impacted by smaller embryo planets and planetesimal size objects. From crater counts it is inferred that a period of intense meteorite impacts began around 4.1 BYA and concluded around 3.8 BYA or at the end of the Hadean eon. In addition, at this time volcanic activity on Earth was severe due to the large heat flow and geothermal gradient on Earth.

By the beginning of the Archean eon at around 3.8 BYA, the Earth had cooled significantly. It is believed that primordial life started to evolve at this time with candidate fossils dating to around 3.5 BYA. Some scientists speculate that life may have started earlier in the Hadean eon.

Earth is often described as having three atmospheres (Fig.1.4). The first atmosphere at the time of Earth's development included elements from solar nebula or molecular hydrogen clouds, mostly hydrogen and helium. A combination of solar winds and Earth's heat would have quickly driven off this primitive atmosphere.

The molten Earth released volatile gases. More gases were released by volcanoes. The gases together created a second atmosphere, an atmosphere of carbon dioxide, carbon monoxide, steam, nitrogen, hydrogen, and ammonia. As

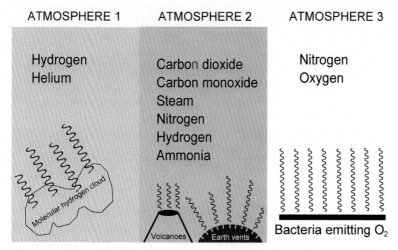

FIGURE 1.4   The three atmospheres on planet Earth.

the planet cooled 4.4 BYA, clouds formed and rain commenced leading to the formation of oceans, seas, rivers, and lakes around the planet. By the start of the Archean eon, 3.8 BYA, the oceans covered much of Earth.

It was not until 2.8 BYA, or until the Earth was 1.7–2.8 billion years old, that a third atmosphere, rich in oxygen, emerged. This occurred when bacteria began to produce oxygen through photosynthesis. Initially, the released oxygen was bound up with limestone, iron, and other minerals. Then, oxidized iron appears as red layers in geological strata called banded iron, that formed in abundance during the Siderian period (between 2.5 and 2.3 BYA). When most of the exposed readily reacting minerals were oxidized, oxygen finally began to accumulate in the atmosphere. Though each photosynthetic cell only produced a small amount of oxygen, the combined metabolism of many cells over a vast time transformed Earth's atmosphere to its current state, Earth's third atmosphere.

When the Earth warmed, it formed a rocky crust. The oldest crust on earth contained rocks rich in calcium and aluminum. The Earth's core is comprised of iron and nickel, while the Earth's mantle or outer layer is mostly comprised of sand-like silicates, magnesium, and iron (Fig. 1.5).

Life began on Earth long before the oxygen containing atmosphere came about. The earliest life forms were minute, and their fossils look like small rods. The oldest undisputed evidence of life was fossilized bacteria dating to 3 BYA, this was during the period when the atmosphere was filled with toxic gas. Other less certain finds in rock, date to around 3.5 BYA; these have also been interpreted as bacteria, and date back to a carbon monoxide toxic gas atmosphere.

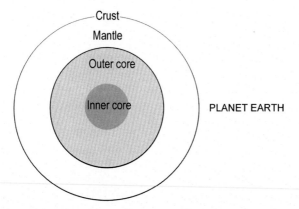

Crust
Oxygen, aluminum,silicon, iron, magnesium, calcium, sodium

Mantle
Silicates, magnesium, iron

Outer core
Iron, nickel

Inner core
Iron, nickel

FIGURE 1.5   The composition of planet Earth.

REFERENCES

[1] Penzias AA, Wilson RW. A measurement of excess antenna temperature at 4080 Mc/s. Astrophys J Lett 1965;142:419–21.
[2] Montmerle T, Augereau J-C, Chaussidon M, et al. Solar system formation and early evolution: the first 100 million years. Earth, Moon, and Planets 2006;98:39–95.
[3] Woolfson MM. Solar system – its origin and evolution. Q J R Astron Soc 1993;34:1–20.
[4] Wiechert U, et al. Oxygen isotopes and the moon-forming giant impact. Science 2001;294:345–8.
[5] Nield T. Moonwalk. Geol Soc London 2008:8.
[6] Wilde SA, Valley JW, Peck WH, Graham CM. Evidence from detrital zircons for the existence of continental crust and oceans on the Earth 4.4 Gyr ago. Nature 2001;409:175–8.

Chapter 2

How Life Started

How life started is a very big question. All answers, including those proposed here are strictly hypothetical. This is because nobody was present 4 billion years ago (BYA) to witness how life started and because there are no first life-forms alive on Earth today.

Yes, the earth had to cool down and form a solid hot mass before life had a chance of starting. That limits us to around 4.2 billion years ago. Furthermore, the atmosphere needed to change from hydrogen and helium to a volcanic atmosphere of carbon dioxide, carbon monoxide, nitrogen, steam, methane, and ammonia before life had a chance to start. That change occurred around 4.4 BYA. The Earth's volcanic onslaught possibly needed to stop as did the bombardment of Earth from space by comets and meteorites before life could take a grasp. They both stopped by around 3.8 BYA.

It is suggested that somewhere on Earth, close to 3.7–4.2 BYA, a set of molecular reactions inside a primitive chamber or cell flicked a switch and became life. Scientists try to imagine this key event by abridging the processes that characterizes living things. New research suggests that this simplification of life needs to go much further.

All known species today rely on DNA or RNA to replicate and reproduce, and proteins to run cellular machinery. These large molecules, DNA, RNA, and proteins, involving thousands of atoms are not likely to have been around for the first organisms to use. Many scientists believe that life started from small molecules.

Scientists hold that the first life-forms were simple chemistry models that somehow grew, reproduced, and even evolved without the need for complex molecules like DNA, RNA, and proteins that define biology as we know it today. How do we define life? Life must be defined as an organism that grows and can in some way duplicate or replicate itself. It may be simple organic chemicals, simple biochemicals, more complex biochemicals, or something as complex as modern day life containing massively complex DNA, RNA, and proteins.

An origin of life story begins with complex biological compounds assembled by sheer chance out of an organic broth on the early Earth's surface. This chance synthesis culminated in biomolecules able to make copies of itself.

The first support for the idea of how life arose out of the primordial soup comes from the famous 1953 experiment by Stanley Miller and Harold Urey [1], in which they made amino acids, the building blocks of proteins, by applying

Biology of Life. http://dx.doi.org/10.1016/B978-0-12-809685-7.00002-2

sparks to a test tube of hydrogen, methane, ammonia, and water, early ingredients of the Earth and its atmosphere [1]. If amino acids could come together out of raw ingredients of the atmosphere, then bigger, more complex molecules could presumably form given enough time. Biologists have devised various situations in which this assemblage takes place in tidal pools, volcanic vents, or on the exterior of clay sediments.

But were the first complex molecules proteins, DNA, or something else? Biologists face the problem of which came first; proteins or enzymes are needed to replicate DNA, but DNA is necessary to coach the building of proteins. Many scientists think that RNA, a cousin of DNA, may have been the first multifaceted molecule on which life was based. RNA can carry genetic information as well as DNA, and it can also control chemical reactions as enzymes do.

Some scientists, however, think this "RNA world" is still too complex to be the origin of life. Information molecules like RNA are polymers of molecular fragments. The primordial soup would be full of things that would terminate these sequences before they grew long enough to be useful. In the very beginning, you couldn't have genetic materials that could copy themselves unless you had chemists showing the polymers what to do.

Instead of complex molecules, life must have started with simple molecules interacting through a cycle of reactions. These reactions would produce compounds that would feedback to the cycle, creating a reaction system. All the interrelated chemistry might be contained in simple membranes, simple fatty acid micelles, or what physicist Freeman Dyson calls "garbage bags." These might have divided just like cells, with each new bag carrying the chemicals to restart or replicate the reaction system. In this way, pseudogenetic information could be passed down. The reaction system could have evolved by creating more complicated molecules like proteins and RNA that would perform the reactions better than the small molecules.

Is it possible that the earliest life-form replicated in a very simple way? It may have divided in two leaving each half with 50% of everything. Each half then had the job of absorbing fresh amino acids and fresh nucleotides to build its contents up to starting levels, before further division occurred. Thus replication was completed.

A candidate for a simple chemical reaction was discovered in 2006 in an undersea microbe, *Methanosarcina acetivorans*, which metabolizes carbon monoxide and expels methane and acetate salts (carbon monoxide–acetate energy system) [2]. In the process, it converts multiple adenosine diphosphate (ADP) molecules to adenosine triphosphate (ATP) [2]. It was found that this primitive organism obtained ATP energy from a reaction with carbon monoxide [2]. Compared to other energy-harnessing processes that require dozens of proteins or enzymes, this acetate-based reaction runs with the help of just two very simple enzymes. Scientists propose that this stripped-eden was what the first organisms used to power their growth. It is possible that this system is where evolution emanated from; it is perhaps, the father of all life.

The oldest fossils found in any structure today, stromatolites, are approximately 3.5 billion years old. These structures are fossils formed by photosynthesizing cyanobacteria. Simpler organisms involving monopeptides and mononucleotides must have appeared much earlier than stromatolites. Stromatolites are like upside-down ice cream cones or egg cartons. New studies suggest that the old stromatolites found in Western Australia may be among the earliest signs of life on the planet.

Stromatolites were discovered in 1956, scientists vary between attributing them to ancient microbes or attributing them to areas of volcanic or volcanic vent activity. In a new study in the June 8 issue of the journal *Nature*, Australian researchers contend that the shapes of the stromatolites are too diverse to have been formed by simplistic processes, or to have played a role in the start of life. Unless stromatolites or cyanobacteria arrived on Earth as part of a comet or meteorite as suggested in the Panspermia Theory.

Abigail Allwood of Macquarie University in Sydney analyzed a stretch of rock formations in Australia and identified seven different types of stromatolites [3]. Separate from ice cream cones and egg cartons, the researchers also found stromatolites that look like fossilized sand dunes or choppy ocean waves that have been frozen and turned to stone [3].

If the cyanobacteria do prove to be the first forms of life, it could change how scientists think about life on early Earth. Many current theories about early life state that the first organisms arose around volcanic vents and other extreme environments. But the Australian cyanobacteria are thought to have formed in normal marine conditions.

If the stromatolites were formed by microbes, then life must have adapted to normal, nonextreme environments early in the planet's history, about 3.5 BYA. Earth is about 4.54 billion years old. Furthermore, life by that time may have already diversified enough from simple chemical life models to form complex systems.

The simple life systems described previously used simple energy pathways and simple reproductive systems that do not exist today, so cannot be confirmed. Without studying them biologically we cannot prove that they existed. Logically they had to exist before 3.5 BYA. The simple life systems must have been ancestors to cyanobacteria and other early species. We have to assume that they existed and developed complexity possibly over 500–700 million years before cyanobacteria or around 4–4.2 BYA.

It is proposed that first life came about by chance in a volcanic environment. Volcanic environments were abundant 4.2–3.8 BYA. Volcanic environments both existed on land and in underwater vents. That is where very simplistic life-forms started with simple chemistry.

It is possible that the earliest life-form started in a microdrop of simple fatty acid in aqueous solution, forming a bilayer vesicle or simplistic cell. This probably existed in a volcanic environment or underwater volcanic vent environment. A fatty acid drop vesicle may have contained a few simplistic amino acids, nucleotides, and random RNA. The diet of this simplistic life-form may have been amino acids and nucleotides absorbed from the volcanic environment. By sheer chance this life-form may have been started by the establishment of the very simplistic carbon monoxide–acetate ATP generating energy system in a carbon monoxide-containing atmosphere [2].

This simple early life-form may have reproduced by the bilayer vesicle bursting into the environment and then forming two smaller vesicles. Each of these sharing an equal distribution of molecules, possibly 50%. Each then grew by the accumulation of additional amino acids, nucleotides, and nutrients from their volcanic diet and divided further. This may have been the first form of life.

As postulated in Figs. 2.1 and 2.2 the advancement of the earliest life-form to a life-form preceding cyanobacteria may have occurred over 10 steps over 500–700 million years. If you consider that primates advanced from early simple prosimian primates with a tiny brain (0.07% of body weight) to lower simian primates (brain 0.19% of body weight), to great apes (brain 0.74% of body weight), and onto hominids (brain 1.2% of body weight) and then humans with a giant brain (2.4% of body weight) over 55 million years. Then approximately 55/5 or 11 million years is required per major step of evolutionary advancement. In this case, simple life may have gone through as many as 64 advancement steps, or six times as many steps as in the model in Figs. 2.1 and 2.2 during 700 million years.

The advancement steps could have been more fractional or more specific rather than those proposed. Translation, very intricate protocols, may have advanced themselves over 11 evolutionary steps rather than the 5 steps indicated in the Figs. 2.1 and 2.2.

In conclusion, it is proposed that life may have started as long ago as 4.2 BYA. Life probably started with a very simple life-form which used amino acids and nucleotides, but not complex proteins and enzymes. The life-form possibly utilized a simplistic carbon monoxide–acetate energy pathway. However, the primitive early life-form was eventually replaced with more advanced life-forms and does not survive in the world today.

Yes, it is possible that life came on an asteroid from elsewhere in the universe (eg, the Panspermia Theory), but there is no evidence to support this concept. If someday we discover life very similar to Earth's life on another planet, this hypothesis may be considered more seriously. Until this time, this hypothesis has to be classed with UFOs as very questionable. So do hypothesis suggesting that life's ingredient may have arrived on Earth from an asteroid or comet and other far-fetched hypotheses. These hypothesisall must be rejected for the time being.

Returning to the earliest life-form and simple life model in Figs. 2.1 and 2.2, the model clearly raises multiple questions regarding its ingredients, how the ingredients expanded, the time frame, or the entire hypothesis. Here we try to explain the entire basis for this book's hypothesis of chemical life and its 11 steps.

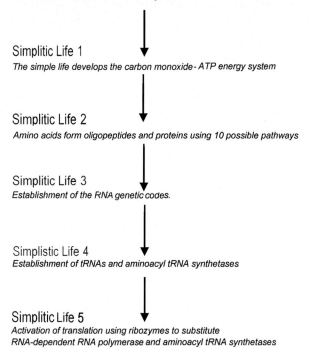

The Earliest Life Form

A drop of fatty acid formed a micelle which led to a bilayer vesicle. Protocells were made in a fatty acid vesicles containing nucleotides, amino acids, oligoribonucleotides (random RNA sequences) and a saline solution.
This was the like the first life form. The protocell had to contain some RNA ribozyme random sequences to aid in early enzyme-like reactions and transport of amino acids and nucleotides into the protocell. The protocells must have used a ribozyme RNA polymerase activity to make RNA from nucleotides which the cell absorbed.
I call this protocell an initial living being. Science evidence shows that vesicles burst spilling their contents into the environment. The vesicle can then reform as multiple smaller vesicles each sharing the initial ingredient. The reformed vesicles absorb the missing ingredients from their environment and makes the needed RNA or ribozymes.

Simplitic Life 1
The simple life develops the carbon monoxide- ATP energy system

Simplitic Life 2
Amino acids form oligopeptides and proteins using 10 possible pathways

Simplitic Life 3
Establishment of the RNA genetic codes.

Simplistic Life 4
Establishment of tRNAs and aminoacyl tRNA synthetases

Simplitic Life 5
Activation of translation using ribozymes to substitute RNA-dependent RNA polymerase and aminoacyl tRNA synthetases

FIGURE 2.1   A postulate linking the earliest life-form with early cyanobacteria.

Firstly, is the time span chosen of 4.2 MYA to 3.5 MYA appropriate? Lazcano and Miller have examined this issue [4], and concluded that it must have taken at least 200 million years for cyanobacteria (stromatolites) to have evolved from the earliest life-form. Here 700 million years or a longer period is proposed, giving room for steps to have evolved slower.

Secondly, why was a volcanic environment selected? The studies of Maher and Stephenson show that a volcanic environment is the best setting for the origin of life to occur, and for nucleotide and amino acid ingredients to be found [5].

Thirdly, why were bilayer vesicles considered for the cells on earliest life-form? How did the vesicles advance to phospholipid membranes? I refer you to the studies of Szostak and colleagues at Massachusetts General Hospital

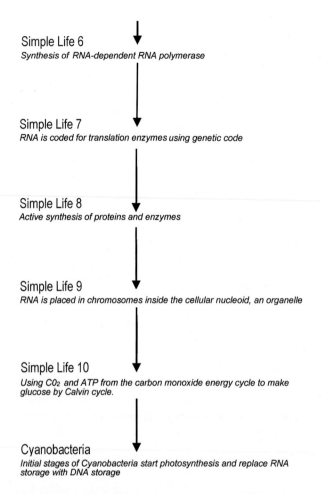

Simple Life 6
Synthesis of RNA-dependent RNA polymerase

Simple Life 7
RNA is coded for translation enzymes using genetic code

Simple Life 8
Active synthesis of proteins and enzymes

Simple Life 9
RNA is placed in chromosomes inside the cellular nucleoid, an organelle

Simple Life 10
Using CO_2 and ATP from the carbon monoxide energy cycle to make glucose by Calvin cycle.

Cyanobacteria
Initial stages of Cyanobacteria start photosynthesis and replace RNA storage with DNA storage

FIGURE 2.2 A postulate linking the earliest life-form with cyanobacteria.

who have conducted multiple studies to address these issues [6,7]. They examine how fatty acid vesicles may have formed, grow, and divide in the early environment. Bilayer vesicles are a step beyond micelles as an interim cell membrane.

The Szostak lab has shown that vesicle formation may also be catalyzed by the clay montmorillonite. Clays such as montmorillonite may very well have been the key to the formation of the first protocells. How do fatty acid vesicles grow? Research in the Szostak lab has shown that when fatty acid micelles are added to a solution of preformed vesicles, the vesicles grow rapidly. Individual fatty acids are transferred from the micelles to the outer leaflet of the vesicle membrane. Fatty acids may then flip from the outer leaflet to the inner leaflet, which allows a membrane bilayer to grow evenly [6,7].

Fourthly, how were amino acids formed in the volcanic atmosphere on primitive Earth? I refer the reader to those key studies completed by Miller and Urey in the 1950s. A variety of experiments were conducted with very basic organic compounds like ammonia, carbon monoxide and carbon dioxide, and methane representative of the early atmosphere and sparks to represent lightning [1,8]. As demonstrated, amino acids and molecules like glucose were readily formed [1,8].

Fifthly, so primitive amino acids were found, how were oligopeptides or small proteins formed? In the 1960s Sidney Fox studied the spontaneous formation of oligopeptides under conditions that existed on early Earth [9]. He demonstrated that in the environment that amino acids attached to each other through carboxyl terminal to amino group, spontaneously forming oligopeptides [9].

Sixthly, how did a complex molecule such as ribonucleotide (a sugar–nucleic acid complex) come about naturally on primitive Earth? I refer you to the studies of John Sutherland at Cambridge University. As shown, simple organic ingredients such as acetylene and formaldehyde present in a volcanic environment could readily undergo a reaction in a volcanic environment to form ribonucleotides [10].

Seventhly, how did ribonucleotides combine together to form oligonucleotides or RNA? The studies of Stan Palasek address this issue, showing that in a volcanic environment as chosen for our model, that ribonucleotides can randomly polymerize to form oligonucleotides or short pieces of RNA [11].

Eighthly and finally, how did RNA become an enzyme and catalyze translation? It has been shown by Cech and by Yarus that it may have been feasible, in early life, for RNA to act as an enzyme [12,13].

This is just a short collection of the articles that support this book's hypothesis on chemical life (Figs. 2.1 and 2.2), dozens more can be found in the literature and on the internet. I welcome your comments and questions about this model of how life started and developed to the level of cyanobacteria. Please email me at: larry@hCGlab.com.

REFERENCES

[1] Miller SL, Urey HC. Organic compound synthesis on the primitive earth. Science 1959;130:245–51.
[2] Ferry JG, House CH. The stepwise evolution of early life driven by energy conservation. Mol Biol Evol 2006;23:1282–92.
[3] Allwood A, Grotzinger K, Anderson B, Coleman K. Controls on development and diversity of Early Archean stromatolites. Proc Natl Acad Sci 2009;106:9548–55.
[4] Lazcano A, Miller SL. How long did it take for life to begin and evolve to cyanobacteria. J Mol Evol 1994;39:546–54.
[5] Maher KA, Stephenson DJ. Impact frustration and the origin of life. Nature 1988;331:612–4.
[6] Zhu TF, Bubin I, Szostak JW. Preparation of fatty acid micelles. Methods Enzymol 2013;533:283–7.
[7] Szostak J.W. Exploring lifes origins: Fatty acids. www.exploringorigins.org/fatty acids. html.

[8] Miller SL. Production of amino acids under possible primitive Earth conditions. Science 1953;117:528–9.

[9] Fox SW. A theory of macromolecular and cellular origins. Nature 1965;205:336–45.

[10] Powner MW, Gerland B, Sutherland JD. Synthesis of activated pyrimidine ribonucleotides prebiotically in plausible conditions. Nature 2009;459:239–42.

[11] Palasek S. Primordial RNA replication and application in PCR technology. Cornell Univ Lib aRXiv 2013;1305:5581–601.

[12] Cech TR. The RNA worlds in context. Cold Spring Harb Perspect Biol 2012;4:a006742.

[13] Yarus M. Getting past the RNA world: the initial Darwinian ancestor. Cold Spring Harb Perspect Biol 2011;3:a003590.

Chapter 3

RNA, Ribozymes, and Early Life Forms

RNA not DNA molecules were the predecessors to all present day life on Earth [1]. It is widely accepted that present day life on Earth descended from an RNA world, although RNA-based life may not have been the very first life to exist [2,3].

RNA stores genetic data like DNA, and catalyzes chemical reactions similarly to enzymes. It may have also played a major step in the evolution of life [2,3]. The RNA world would have inevitably been replaced by the DNA and enzyme world of today, likely through ribonucleoprotein enzymes such as the ribosome and ribozymes, since enzymes large enough to self-fold and have activities would only have come about after RNA was obtainable to catalyze peptide ligation or amino acid polymerization [2]. DNA is supposed to have taken over the role of information storage because of its increased stability, while enzymes replaced RNA's role as specialized catalysts.

The RNA hypothesis presented here is supported by many lines of evidence, such as the observations that RNA is central to the translation pathway and that small RNAs can catalyze all of the chemical group and data transfers required for life [2,3]. The structure of the ribosome shows that the ribosome is a ribozyme, with a core of RNA and no amino acid side chains and an active site where peptide bond formation is catalyzed. Many of the most acute components of cells are composed mostly of RNA. Also, many critical cofactors (eg, adenosine triphosphate and nicotinamide adenine dinucleotide) are either nucleotides or materials clearly related to RNA. This indicates that RNA in present day cells is an evolutionary relic of an RNA-based enzymatic system that heralded the protein-based system seen in all present life.

The properties of RNA make the RNA world hypothesis reasonable, though its general acceptance as an account for the origin of life requires further research [4]. RNA is recognized to form efficient catalysts and its resemblance to DNA makes its ability to store information clear. Thoughts differ, however, as to whether RNA constituted the first self-replicating system or was a copied from a still-earlier system [2]. One version of the thesis is that an alternative type of nucleic acid, preRNA, was the first to emerge as a self-reproducing molecule, to be replaced later by RNA. On the other hand, the finding that pyrimidine ribonucleotides can be synthesized under

Biology of Life. http://dx.doi.org/10.1016/B978-0-12-809685-7.00003-4

plausible early life conditions [5] means that it is too early to dismiss the RNA-first scenarios [2]. Suggestions for simple preRNA nucleic acids have included, Peptide nucleic acid, Threose nucleic acid, or Glycol nucleic acid [6]. Regardless of their structural simplicity and possession of features comparable with RNA, the generation of simpler nucleic acids has yet to be demonstrated.

Ribozymes, are found in DNA-based life today and could be examples of living fossils. Ribozymes play roles like those in the ribosome, essential for protein synthesis. Other ribozyme functions exist; such as the hammerhead ribozyme performs self-cleavage [7], and an RNA polymerase ribozyme can synthesize short RNA strands from a primed RNA template.

Among the enzymatic properties important for the beginning of life are self-replication. The ability to self-replicate, or synthesize other RNA molecules. Relatively short RNA molecules that can synthesize others have been artificially produced in the laboratory. The shortest was 165 bases long. One version that was 189 bases long, had an error rate of just 1.1% per nucleotide when synthesizing an 11 nucleotide long RNA strand from primed template strands [8]. The longest primer extension performed by a ribozyme polymerase was 20 bases.

The capacity to catalyze simple chemical reactions which enhanced creation of molecules that are building blocks of RNA molecules. Short RNA molecules with such abilities have been artificially formed in the laboratory. A recent study showed that almost any nucleic acid can evolve into a catalytic sequence under appropriate selection. Amino acid-RNA ligation. The ability to conjugate an amino acid to the 3'-end of an RNA in order to use its chemical groups or provide a long-branched aliphatic side-chain.

The talent to catalyze the formation of peptide bonds to produce short peptides or longer proteins, is achieved in present day cells by ribosomes. A short RNA molecule has been created in the laboratory with the ability to form peptide bonds, and it has been suggested that rRNA has evolved from a similar molecule [9]. It has also been proposed that amino acids may have initially been involved with RNA molecules as cofactors enhancing or diversifying their enzymatic capabilities, before evolving to more complex peptides. Likewise, tRNA is suggested to have evolved from RNA molecules that began to catalyze amino acid transfer.

RNA is a similar molecule to DNA (Fig. 3.1), and has only two chemical differences from DNA. One strand of DNA and one of RNA can bind to form a double helical structure. This makes the loading of information in RNA possible in a very similar way to DNA, however RNA is notably less stable. The major difference between RNA and DNA is the presence of a hydroxyl group at the 2'-position on the pentose sugar (Fig. 3.1).

RNA also uses different bases than DNA. RNA uses adenine, guanine, cytosine, and uracil, while DNA uses adenine, guanine, cytosine, and thymine. Chemically, uracil is similar to thymine, differing only by a single methyl group; this has no affect on base pairing as adenine easily binds uracil or thymine. RNA is thought to have come before DNA, because of their ordering in the biosynthetic pathways. The deoxyribonucleotides used to make DNA are actually made from ribonucleotides.

Riboswitches have been shown to act as regulators of gene expression, most notably in bacteria, but also in plants. Riboswitches alter their secondary structure in response to the binding of a metabolite. This change in structure can result in the formation or disruption of a terminator, truncating or permitting transcription respectively. It has been suggested that riboswitches originated in

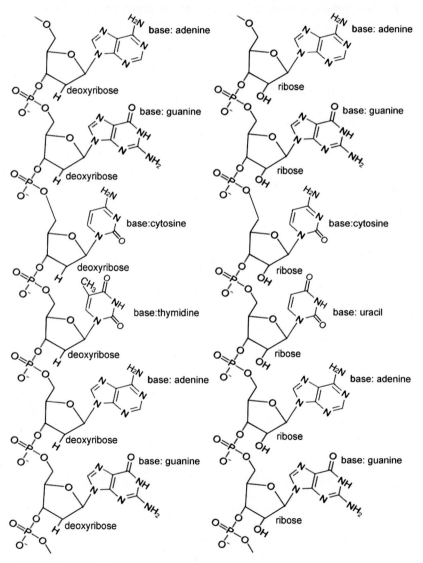

FIGURE 3.1 Comparison of a hexanucleotide of DNA (left) and RNA (right) showing bases.

an RNA-based world. In addition, RNA thermometers regulate gene expression in response to temperature changes [10].

Nucleotides are the fundamental molecules that combine together to form RNA. They consist of a nucleic acid base attached to a sugar-phosphate backbone. RNA is made of long sections of specific nucleotides arranged so that their bases carry specific information. The RNA world hypothesis states that in the primordial soup, there existed free nucleotides. These nucleotides formed bonds with one another. These chains have been proposed by some as the first, primitive forms of life [11]. As the principal sets of RNA molecules expanded their numbers, novel catalytic properties added by mutation, which benefitted their persistence and expansion, could accrue in the population.

Rivalry may have chosen the emergence of cooperation between different RNA chains, opening the way for the formation of the first protocell. RNA chains developed with catalytic properties that helped peptide bonding. These peptide-bonded amino acids could then assist with RNA synthesis, giving those RNA chains that could serve as ribozymes the selective advantage. The facility to catalyze one step in protein synthesis, aminoacylation of RNA, has been shown in a short five-nucleotide segment of RNA [12].

The RNA world hypothesis, if true, has essential implications for the definition of life. For most of the time that followed Watson and Crick's clarification of DNA structure in 1953, life was mainly defined in terms of DNA and proteins. DNA and proteins seemed the dominant macromolecules in the living cell, with RNA helping in creating proteins from the DNA blueprint.

Other interesting discoveries demonstrated a role for RNA beyond a simple message or transfer molecule. These include the importance of small nuclear ribonucleoproteins in the processing of premRNA and RNA editing, RNA interference, and reverse transcription from RNA in eukaryotes in the maintenance of telomeres in the telomerase reaction.

REFERENCES

[1] Zimmer CA. Tiny emissary from the ancient past. New York Times; September 25, 2014.
[2] Cech TR. The RNA worlds in context. Cold Spring Harb Perspect Biol 2012;4:a006742.
[3] Yarus M. Getting past the RNA world: the initial Darwinian ancestor. Cold Spring Harb Perspect Biol 2011;3:a003590.
[4] Atkins JF, Gesteland RF, Cech T, Thomas R. The RNA world: the nature of modern RNA suggests a prebiotic RNA world. Cold Spring Harbor Laboratory Press; 2006.
[5] Powner MW, Gerland B, Sutherland JD. Synthesis of activated pyrimidine ribonucleotides prebiotically in plausible conditions. Nature 2009;459:239–42.
[6] Orgel LA. Simpler nucleic acid. Science 2000;290:1306–7.
[7] Forster AC, Symons RH. Self-cleavage of plus and minus RNAs of a virusoid and a structural model for the active sites. Cell 1987;49:211–20.
[8] Johnston WK, Unrau PJ, Lawrence MS, Glasner ME, Bartel DP. RNA-catalyzed RNA polymerization: accurate and general RNA-templated primer extension. Science 2001;292:1319–25.

[9]   Zhang B, Cech TR. Peptide bond formation by in vitro selected ribozymes. Nature 1997;390:
       96–100.
[10]   Narberhaus F, Waldminghaus T, Chowdhury S. RNA thermometers. FEMS Microbiol Rev
       2006;30:3–16.
[11]   Villarreal LP, Witzany G. The DNA habitat and its RNA inhabitants: at the dawn of RNA
       sociology. Genomic Insights 2013;6:1–12.
[12]   Turk RM, Chumachenko NV, Yarus M. Multiple translational products from a five-nucleotide
       ribozyme. Proc Natl Acad Sci USA 2010;107(10):4585–9.

Chapter 4

Amino Acids and Peptides, Essential Ingredients of Life

In addition to ribonucleotides, amino acids must be considered as an essential ingredient in life. We consider here how amino acids were formed in early life, how peptides and small proteins were formed, and how essential they are to life. Proteins and peptides function as: hundreds of enzymes; building blocks for the structural features of tissues and muscles; uptake acceptors and signal receptors in cells; hormones; and other features of life.

Firstly, we consider amino acid synthesis under very primitive earth conditions. It is essential to first consider the Miller–Urey keynote experiments [1,2]. The Miller–Urey experiments were investigations that tested the chemical origin of life ingredients under primitive Earth conditions. Specifically, the experiments investigated Alexander Oparin's 1924 hypothesis that conditions on the primitive Earth favored chemical reactions that synthesized essential life building blocks from simpler inorganic precursors [3]. Trying to mimic the early Earth's environment with ingredients and lightning. The Miller–Urey experiment is illustrated in Fig. 4.1.

After the death of Miller in 2007, scientists tested sealed vials of "organic compound in water" preserved from the original Miller–Urey experiments and were able to show that there were actually well over 20 different amino acids produced in the original experiments (Fig. 4.1). That is much more than what Miller reported. All 20 amino acids were found that naturally occur in life.

There is abundant evidence that the major volcanic eruptions that occurred four billion years ago would have released CO, CO_2, N_2, and SO_2 into the atmosphere. Studies using the same experiment with these gases in addition to the Miller experiment four gases would have produced many more diverse molecules [4]. Recent studies show that Earth's original atmosphere might have had an alternative composition from the gases used in the Miller experiment. But studies continue to show mixtures of simple to complex compounds in the Miller–Urey experimental protocol under varying conditions (Fig. 4.1) [5].

Other studies by Ferris and colleagues [6] showed that amino acids can be formed by HCN treatment of simple organics such as N-alkyliminoacetonitriles. HCN and simple organics were surely abundant in the primitive world. These experiment explain how and why amino acids were found in the primitive

Biology of Life. http://dx.doi.org/10.1016/B978-0-12-809685-7.00004-6

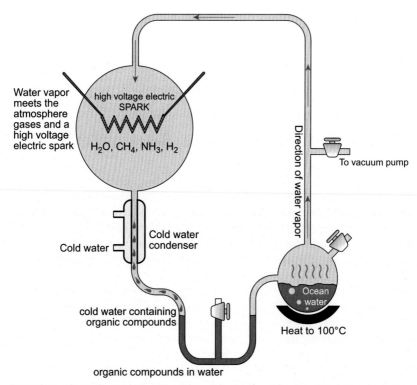

FIGURE 4.1 The Miller–Urey Experiment. Electric flashes in an H_2O, CH_4, NH_3, and H_2 atmosphere.

world. Others have postulated that amino acids may have arrived on Earth on comets as in the Panspermia theory. However, the Miller–Urey studies have proven that all 20 amino acids seem to occur on the Earth naturally.

The next question is how did these amino acids polymerize and form proteins and peptides on primitive Earth? Investigators have now proposed nine different ways that this may have occurred on Earth and in volcanic areas on Earth. It is my opinion that several of the proposed peptide polymerizations methods are invalid, use of extreme heat (175°C); did such heat really occur in such a controlled manner? Condensation of amino acid in 160°C ammonia is that really practical?

Investigators have proposed several ways in which proteins and peptides may have been formed, these include:

High energy radiation and ion polycondensation which polymerized amino acids [7].

Controlled exposure to heat at 175°C causing amino acids to polymerize into proteins [8].

Salt induced peptide polymerization with the assistance of clay materials [9].

Condensation of amino acids in ammonia at 160°C, which generated polymers [10].

Creation of volcanic settings in which amino acids were converted into peptides by the use of coprecipitated (Ni, Fe)S and CO in conjunction with H_2S [11].

Carbon disulfide, a gas discharged from volcanoes, which causes homo- and hetero-peptides to be produced [12].

HCN and other ingredients of primitive atmospheres can polymerize amino acids [13].

Heated ammonium cyanide which causes amino acids to polymerize [14].

Clay and water subjected to cyclic variations in temperature formed oligopeptides [15].

I am not sure which of these nine conditions existed on the early Earth, or what combinations of these conditions occurred but somehow amino acids became polymerized to form random proteins. How these early random proteins became useful is another question. Surely it would have required a million proteins to be randomly made before one was made that had a function. I refer to the research of van der Gulik and colleagues [16] to answer this question.

The research of van der Gulik and colleagues attempts to tackle the question of what did the first peptides look like when they emerged on the primitive earth, and what simple catalytic activities did they fulfill [16].

They guess that the early peptides were short, three to eight amino acids long, were made mostly of the simple amino acids Gly, Ala, Val, and Asp, that are abundantly produced in many prebiotic synthesis experiments, and that the neutralization of Asp's negative charge is achieved by metal cations. They further assume that some traces of these prebiotic peptides still exist, in the form of active sites in present-day proteins. Searching proteins for prebiotic peptide candidates led them to identify three classes of peptides: Asp–Tyr–Asp–Gly–Asp corresponding to the active site in RNA polymerases, Asp–Gly–Asp–Ala–Asp is present in some kinds of mutases, and Asp–Ala–Lys–Val–Gly–Asp–Gly–Asp is present in dihydroxyacetone kinase. All three peptides contain an Asp–Gly–Asp subgroup, which is suggested to be the common ancestor of all active peptides. Moreover, all three manipulate phosphate groups, which was probably a very important biological function in the very first stages of life. The significance of these results is supported by the frequency of these motifs in today's proteins, which is three times higher than expected by chance, with a P-value of 3×10^{-2}.

REFERENCES

[1] Miller SL. Production of amino acids under possible primitive earth conditions. Science 1953;117:528–9.

[2] Miller SL, Urey HC. Organic compound synthesis on the primitive earth. Science 1959;130:245–51.

[3] Oparin AI. The origin of life. 1936. Moscow.

[4] Cleaves HJ, Chalmers JH, Lazcano A, Miller SL, Bada JL. A reassessment of prebiotic organic synthesis in neutral planetary atmospheres. Origins Life Evol Biosph 2008;38:105–15.

[5] Bada JL. New insights into prebiotic chemistry from Stanley Miller's spark discharge experiments. Chem Soc Rev 2013;42:2186–98.

[6] Ferris JP, Donner DB, Lotz W. The possible role of HCN in the origins of life. The oligomerization of HCN and the reaction of HCN with N-alkyliminoacetonitriles. Bioorg Chem 1972;2:95–8.

[7] Mathews CN, Mosner RE. Prebiological protein synthesis. Proc Natl Acad Sci USA 1966;56:1087–94.

[8] Fox SW, Harada K. Thermal copolymerization of amino acids to a product resembling protein. Science November 14, 1958;1958(14):128.

[9] Rode BM. Peptides and the origins of life. Peptides 1999;20:773–86.

[10] Guidry CL. Direct synthesis of polypeptides. I. Polycondensation of glycine in aqueous ammonia. Arch Biochem Biophys 1961;93:166–71.

[11] Huber C, Wächtershäuser G. Peptides by activation of amino acids with CO on (Ni,Fe)S surfaces: implications for the origin of life. Science 1998;281:670–2.

[12] Lehman L, Orgel L, Ghadiri MR. Carbonyl sulfide-mediated prebiotic formation of peptides. Science 2004;306:283–6.

[13] Oro J, Kamat SS. Amino-acid synthesis from hydrogen cyanide under possible primitive earth conditions. Nature 1961;190:442–3.

[14] Lowe CU, Rees MW, Markham R. Synthesis of complex organic compounds from simple precursors: formation of amino-acids, amino-acid polymers, fatty acids and purines from ammonium cyanide. Nature 1963;199:219–26.

[15] Lahav N, White D, Chang S. Peptide formation in the prebiotic era: thermal condensation of glycine in fluctuating clay environments. Science 1978;201:67–9.

[16] van der Gulik P, Massar S, Gilis D, Buhrman H, Rooman M. The first peptides – the evolutionary transition between prebiotic amino acids and early proteins. J Theor Biol 2009;261:531–9.

Chapter 5

Biochemistry of Early Life

When I contracted this book writing project, it was completely new to my mind. I locked myself away in my office and read, reread, and reread once again 53 articles on how early biochemical molecules came about, about how early cells were formed, about early cell energy systems, and about the early demonstration of life. These articles are all listed in the references to this chapter.

The Introduction to this book describes my own strange brain and how it works. Somehow, these 53 references of data became stored in the science side or left side of my brain, jumbled up with 37 years of human chorionic gonadotropin data, gynecology and obstetrics data, and detailed biochemistry knowledge. My left side somehow churned this data over for three long weeks.

Asking myself the question, how did life first start and how did life first begin? I wrote down what my brain considered a logical theory to explain the origin of life and how life may have developed from ~4.2–4 BYA to cyanobacteria by 3.5 BYA.

I first considered the timetable of when life had to first start. Planet Earth was made 4.54 BYA. Today, the oldest fossils that can be found are stromatolites from cyanobacteria 3.5 BYA. Cyanobacteria are quite sophisticated lifeforms, which utilize photosynthesis, RNA, and oxidative phosphorylation. This strongly indicated that simple life had to have started in the hundreds of million years prior to cyanobacteria [1–4]. Seemingly, simple life-forms left no fossils or other evidence on the Earth today. The Earth had to cool down from its creation to permit life-forms to exist, that limits us to about 4.2 BYA, or to 4.2–3.5 BYA for simple life to have started [1–4].

At that time, the atmosphere on Earth was very different to the oxygen–nitrogen atmosphere of today; the atmosphere was comprised of CO, CO_2, N_2, and SO_2. This atmosphere was essential for amino acid synthesis, and triggered many proposed pathways [5]. Volcanic activity was rampant and volcanic regions such as, undersea volcanic vents, geysers, and areas near active volcanoes were optimal sites for the chemicals to begin life [6–8].

I selected 53 articles to pay to have printed. 20 were the most cited articles in the origin of life field; 26 were key articles cited by the 20 most cited articles; and 7 articles were found by me to answer specific questions. From the 53 articles studied, it is clear that the presence of cell membranes that can transport water-based and ionic molecules was the first requirement of life [9–20]. Initially, fats were formed at low concentrations from CO and H_2 in the atmosphere, particularly in volcanic environments [6–20]. Fats must have concentrated with heat and time. Clearly, volcanic sites are the most likely sites for life to start [6–8]. Fats formed micelles which developed into vesicles and vesicles with a hydrophobic tail which formed a fatty acid lipid bilayer. These were clearly the optimal circular bilayer structure and the starting material for protocell formation [9–16].

Protocells were composed of a fatty acid vesicle containing nucleotides, amino acids, oligoribonucleotides (random RNA sequences) [8,10], and a saline solution. The RNA was seemingly generated in the volcanic environment [8,10]; this was the first life-form. The protocell had to contain some RNA ribozyme random sequences or absorbed intact RNA to aid in early enzyme-like reactions and to transport amino acids and nucleotides into the protocell [21–26]. The protocells might have used a ribozyme RNA polymerase activity to make RNA from nucleotides which the cell absorbed [21–32].

I call this protocell an initial living being. Yet, does it replicate and grow? The protocell is made of a fatty acid bilayer vesicle. Scientific evidence shows that vesicles burst spilling their contents into the environment. The vesicle can then reform possibly as multiple small vesicles each sharing the initial ingredients. The reformed vesicles absorb the missing ingredients from their environment and make the needed RNA or ribozymes. Through these processes, if successful, they effectively reproduced (Fig. 2.1) [10,12,16].

The starting original life-form develops or evolves with time forming Simple Life-1 (Fig. 2.1). Simple Life-1 gains the simplest energy system, the carbon monoxide–Methane/adenosine triphosphate (ATP) energy system. This uses carbon monoxide, expels methane and generates ATP. This pathway requires two enzymes, which can be provided by random ribozymes [21–26] (Fig. 5.1). The carbon monoxide–Methane/ATP energy system generated ATP. This permitted high energy ATP, cytidine triphosphate (CTP), uridine triphosphate (UTP), and guanosine triphosphate (GTP) generation, and allowed ATP to power future enzymatic and present ribozyme reactions.

The second evolutionary step (Simple Life-2, Fig. 2.1) involved polymerization of amino acids to make oligopeptides and proteins. Ten independent approaches have been proposed [33–42]:

- High energy radiation and ion polycondensation may have polymerized amino acids [33].
- Controlled exposure to heat at 175°C caused amino acids to polymerize to proteins [34].

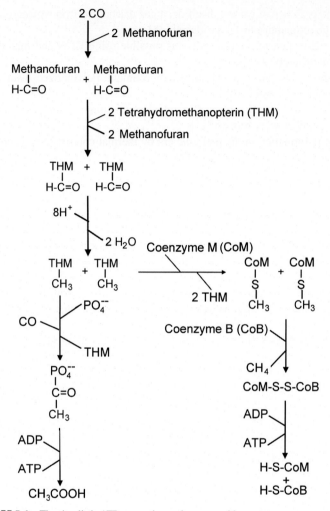

FIGURE 5.1 The simplistic ATP-generating carbon monoxide–acetate energy system [2].

- Salt induced peptide polymerization with the assistance of clay materials [35].
- Condensation of amino acids in ammonia at 160°C generated polymers [36].
- In volcanic settings, amino acids were converted into their peptides by the use of coprecipitated (Ni, Fe)S and CO in conjunction with H_2S [37].

- In the presence of carbon disulfide, a gas discharged from volcanoes, homo- and hetero-peptides were produced [38].
- Polymerization of amino acids was possible using HCN and ingredients of primitive atmospheres [39].
- Amino acids from a primitive environment polymerized using heated ammonium cyanide [40].
- In clay and water subjected to cyclic variations in temperature, longer oligopeptides were formed [41].
- Peptide formation in the prebiotic era by thermal condensation of glycine in fluctuating clay environments [42].

We consider that a combination of all these approaches may have generated oligopeptides and proteins in the proposed model (Figs. 2.1 and 2.2).

The third evolutionary step (Simple Life-3→Simple Life-7, Fig. 2.1) is a multistep evolutionary event involving the activation of RNA-dependent RNA translation in the simple life-form. Firstly, Simple Life-3, is the determination of the genetic code; secondly, Simple Life-4, is the synthesis of amino acids tRNAs; thirdly, Simple Life-5, is the activation of translation using ribozymes to substitute for the specific aminoacyl tRNA synthetases; fourthly, Simple Life-6, is the synthesis of aminoacyl tRNA synthetases; and finally, Simple Life-7, RNA is coded for the synthesized proteins [43].

In Simple Life-8 (Fig. 2.2), active synthesis of amino acids and proteins occurs. In Simple Life-9 (Fig. 2.2), RNA is placed in chromosomes inside the cellular nucleoid (an organelle). In Simple Life-10, cells make enzymes and use them and ribozymes for the Calvin cycle, using CO_2 and ATP from the carbon monoxide energy cycle to make glucose (Fig. 5.2).

Cells then become cyanobacteria, adopting photosynthesis and replacing RNA coding with DNA. Additional references describe sugar synthesis [44], energy pathways [45–47], amino acid synthesis [28,48], and RNA synthesis [49].

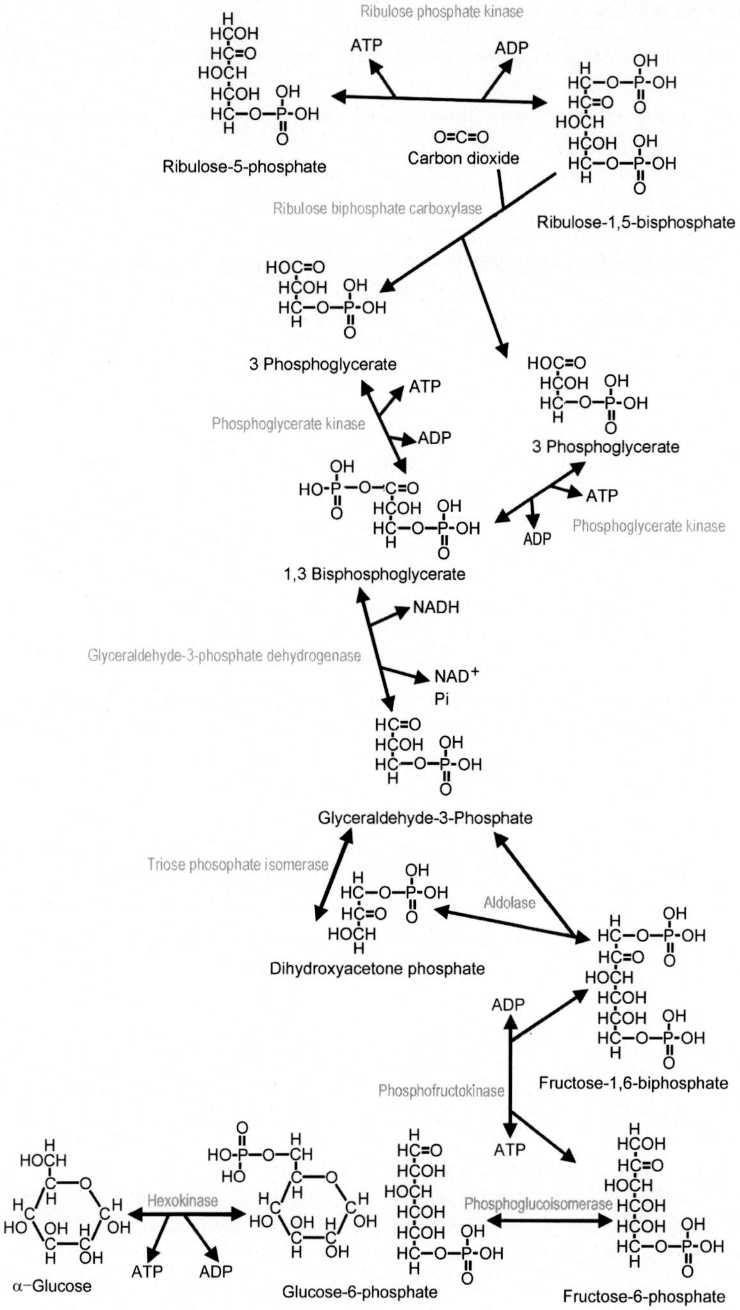

FIGURE 5.2 Calvin cycle and glycolysis for converting CO_2 to glucose.

REFERENCES

[1] Lazcano A, Miller SL. How long did it take for life to begin and evolve to cyanobacteria. J Mol Evol 1994;39:546–54.

[2] Timetable of evolutionary history of life, https://en.wikipedia.org/wiki/timetable_of_the_ evolutionary_history_of_life.

[3] Walsh B. Origins of life, nitro.biosci/Arizona.edu/course/EEB105/lectures/origin_of_life/ origins.htm.

[4] BBC Nature history of life, www.bbc.co.uk/Nature/History.

[5] Zahnle K, Schaefer L, Fegley B. Earth's earliest atmospheres. Deamer D, Szostak J, editors. The Origin of Life. Cold Spring Harbor Perspect Biol July 2015.

[6] Maher KA, Stephenson DJ. Impact frustration and the origin of life. Nature 1988;331:612–4.

[7] Gish D. Origin of life: The Fox thermal model of the origin of life, http://www.icr.org/article/life-fox-thermal-model-origin-life/.

[8] Schirber M. The volcanic origin of life. 2004. http://www.livescience.com/24-volcanic-origin-life.html.

[9] Zhu TF, Bubin I, Szostak JW. Preparation of fatty acid micelles. Methods Enzymol 2013;533:283–7.

[10] Palasek S. Primordial RNA replication and application in PCR technology. Cornell Univ Lib aRXiv 2013;1305:5581–601.

[11] Loeb J. The dynamics of living matter. In: Osborn HF, Wilson EB, editors. Columbia univ biolog sci. vol. VIII. New, York: Columbia Univ Press; 1906.

[12] Schopf JN, Kudryavtsev AB, Czata AD, Tripathi AB. Evidence of Archean life: Stromatolite and microfossils. Precamb Res 2007;158:141–55.

[13] Griffiths G. Cell evolution and the problems of membrane topology. Nature Rev 2007;8:1018–24.

[14] Mast SS, Schrum JP, Krishnamurthy M, Tobe S, Treco DA, Szoestak JW. Template-directed synthesis of a genetic polymer in a model protocell. Nature 2008;454:122–5.

[15] Marchant J. Oil droplets mimic early life. Nature 2011. News 23 February 2011. Fatty acids form vesicles, these can be fatty acid bilayers.

[16] Hanczyc MM, Toyota T, Ikegami T, Packard N, Sugawara T. Fatty acid chemistry at the oil-water interface: self-propelled oil droplets. J Am Cancer Soc 2007;129:9386–91.

[17] Hanczyc MM, Parrilla JM, Nicholson A, Yanev K, Stoy K. Creating and maintaining chemical artificial life by robotic symbiosis. Artif Life 2014;21:47–54. Artificial life created in a fat droplet.

[18] Schrum JP, Zhu TF, Szostak JW. The origins of cellular life. Cold Spring Harbor Perspect Biol 2010:a002212. Vesicles composed of fatty acids form procells; Life is process based on vesicle.

[19] Chen IA, Walde P. From self-assembled vesicles to protocells. Cold Spring Harbor Perspect Biol 2010:a002170.

[20] Douliez J-P, Gaillard C. Self-assembly of fatty acids: from foams to protocell vesicles. New J Chem 2014;38:5142–8.

[21] The RNA and ribozymes chapter, Cole LA (in press).

[22] Cech TR. The RNA worlds in context. Cold Spring Harbor Perspect Biol 2012;4:a006742.

[23] Yarus M. Getting past the RNA world: the initial Darwinian ancestor. Cold Spring Harbor Perspect Biol 2011;3:a003590.

[24] Atkins JF, Gesteland RF, Cech T, Thomas. The RNA world: the nature of modern RNA suggests a prebiotic RNA world. Cold Spring Harbor Laboratory Press; 2006.

[25] Orgel LA. Simpler nucleic acid. Science 2000;290:1306–7.

[26] Johnston WK, Unrau PJ, Lawrence MS, Glasner ME, Bartel DP. RNA-catalyzed RNA polymerization: accurate and general RNA-templated primer extension. Science 2001;292:1319–25.

[27] Powner MW, Gerland B, Sutherland JD. Synthesis of activated pyrimidine ribonucleotides prebiotically in plausible conditions. Nature 2009;459:239–42.

[28] Ferris JP, Donner DB, Lotz W. The possible role of HCN in the origins of life. The oligomerization of HCN and the reaction of HCN with N-alkyliminoacetonitriles. Bioorg Chem 1972;2:95–8.

[29] Senanayake SD, Idriss H. Photocatalysis and the origin of life: synthesis of nucleoside bases from formamide on TiO_2 single surfaces. Proc Natl Acad Sci USA 2006;103:1194–8.

[30] Miller SL, Urey HC. Organic compound synthesis on the primitive earth. Science 1959;130:245–51.

[31] Ferris JP. Montmorillonite-catalysed formation of RNA oligomers: the possible role of catalysis in the origins of life. Philos Trans R Soc Lond B Biol Sci 2006;261:1777–86.

[32] Orgel LA. Prebiotic chemistry and the origin of the RNA world. Crit Rev Biochem Mol Biol 2004;39:99–123.

[33] Mathews CN, Mosner RE. Prebiological protein synthesis. Proc Natl Acad Sci USA 1966;56:1087–94.

[34] Durante M, Gialanella G, Pugliese M, Scampoli P, Furusawa Y, Kanai T, et al. Chromosomal aberrations induced by high-energy iron ions with shielding. Adv Space Res 2004;34:1358–61.

[35] Fox SW, Harada K. Thermal copolymerization of amino acids to a product resembling protein. Science November 14, 1958;1958(14):128.

[36] Rode BM. Peptides and the origins of life. Peptides 1999;20:773–86.

[37] Guidry CL. Direct synthesis of polypeptides. I. Polycondensation of glycine in aqueous ammonia. Arch Biochem Biophys 1961;93:166–71.

[38] Huber C, Wächtershäuser G. Peptides by activation of amino acids with CO on (Ni,Fe)S surfaces: implications for the origin of life. Science 1998;281:670–2. In modeling volcanic settings amino acids were converted into their peptides by use of coprecipitated (Ni, Fe)S and CO in conjunction with H_2S.

[39] Lehman L, Orgel L, Ghadiri MR. Carbonyl sulfide-mediated prebiotic formation of peptides. Science 2004;306:283–6.

[40] Oro J, Kamat SS. Amino-acid synthesis from hydrogen cyanide under possible primitive earth conditions. Nature 1961;190:442–3.

[41] Lowe CU, Rees MW, Markham R. Synthesis of complex organic compounds from simple precursors: formation of amino-acids, amino-acid polymers, fatty acids and purines from ammonium cyanide. Nature 1963;199:219–26.

[42] Lahav N, White D, Chang S. Peptide formation in the prebiotic era: thermal condensation of glycine in fluctuating clay environments. Science 1978;201:67–9.

[43] Van der Gulik P, Massar S, Gilis D, Buhrman H, Rooman M. The first peptides – the evolutionary transition between prebiotic amino acids and early proteins. J Theor Biol 2009;261:531–9.

[44] Burroughs L, Clarke PA, Farinos H, Glks JAR, Hayes CJ, Vale ME, et al. Asymmetric formation of protected and unprotected tetroses under potentially prebiotic conditions. Org Biomol Chem 2012;10:1565–70.

[45] Ferry JG, House CH. The stepwise evolution of early life driven by energy conservation. Mol Biol Evol 2006;23:1282–92.

[46] Fox RF. Origin of life and energy. 2nd ed. W.H. Freeman & Co.; 1988. Origin of the carbon monoxide and oxidative phosphorylation energy systems.

[47] Deamer D, Szostak JW. The origin of life. Cold Spring Harbor Perspect Biol July 2015.

[48] Miller SL. Production of amino acids under possible primitive earth conditions. Science 1953;117:528–30.

[49] Martin WF, Sousa FL, Lane N. Energy at life's origin. Science 2014;344:1092–3.

Chapter 6

Evolutionary History of Planet Earth

OXYGEN ATMOSPHERE

Table 6.1 and Fig. 6.1 illustrate the evolutionary history of planet Earth. By far the biggest event in the Earth's history is the change from a toxic carbon dioxide, carbon monoxide, methane, nitrogen, steam, methane, and ammonia atmosphere to an oxygen- and nitrogen-based atmosphere. This occurred as a consequence of oxygen produced during photosynthesis. Most plants, algae, and water-based cyanobacteria performed photosynthesis; such organisms are called photoautotrophs [1,2]. It is thought that most of the oxygen from photosynthesis for the toxic atmosphere to oxygen-based atmosphere transition came from water-based cyanobacteria (Fig. 6.1). While the early toxic atmosphere aided life's formation, the oxygen-based atmosphere aided life's advancement [1,2].

Fig. 6.2 shows the distribution of energy pathways over the history of planet Earth. Life probably started with primitive species using the carbon monoxide–acetate adenosine triphosphate (ATP)-generating pathway, then with the establishment of photosynthesis this energy pathway became prominent among prokaryotes. Following the change to an oxygen-based atmosphere, the oxidative phosphorylation pathway became prominent among the emerging eukaryotes [1,2].

PROKARYOTIC AND EUKARYOTIC CELLS

The second biggest step in evolutionary history was probably the establishment of eukaryotes. There are two types of cells among all creatures, prokaryotes and eukaryotes (Fig. 6.3). Eukaryotes include single cell species, fungi, plants, all mammals, animals, fish, reptiles, birds, and amphibians. Prokaryotes include bacteria and archaea. Only prokaryotes existed until approximately 1.3 billion years ago when eukaryotes emerged.

In eukaryotes, the plasma membrane of the cell wall is made of a phospholipid bilayer, with charge on the bilayer outer walls and the hydrophobic tails between the two layers (Fig. 6.3). The nucleus is the central site that houses

Biology of Life. http://dx.doi.org/10.1016/B978-0-12-809685-7.00006-X

TABLE 6.1 Evolutionary History of Planet Earth

13,798 MYA	The Big Bang theory of the origin of the universe
4600 MYA	The formation of the sun and the solar system
4540 MYA	The formation of Earth from asteroids and embryo planets
3500 MYA	Earliest evidence of life, cyanobacteria stromatolites
3000 MYA	Evidence of cyanobacteria photosynthesis
2320–2450 MYA	Cyanobacteria (blue-green algae) produce oxygen in atmosphere
1,300,000	First eukaryotes with a nucleus detected, single cells organisms
1200 MYA	Multicellular organisms detected
665 MYA	Invertebrates detected with head
543 MYA	First animals detected, marine invertebrates
540 MYA	Sponges, the first multicellular organism with diversity detected
485 MYA	First appearance of vertebrates and fish, jawless fish
385 MYA	Evolution of first trees, Wattieza
370 MYA	First amphibians, lobe finned fish
363 MYA	First mammals, terrestrial vertebrates, amniotes with lungs and limbs
315 MYA	First reptiles, casineria
65-230 MYA	Reign of dinosaurs
200 MYA	Evidence of cells with nuclei or eukaryotes
160 MYA	First birds, archaeornis huxley
66 MYA	The Cretaceous–Paleogene extinction eradicates half of all animals
0.20 MYA	Modern humans detected

MYA, Millions of years ago.

the chromosomes, or bundles of DNA strands, which are wrapped by histone proteins [3]. The nucleolus is the largest structure in the nucleus of eukaryotic cells, where it primarily serves as the site of ribosome synthesis and assembly. Nucleoli are made of proteins and RNA and form around specific chromosomal regions [3].

The rough endoplasmic reticulum in eukaryotic cells descends out of the nucleus (Fig. 6.3). It is the site of mRNA translation or protein synthesis,

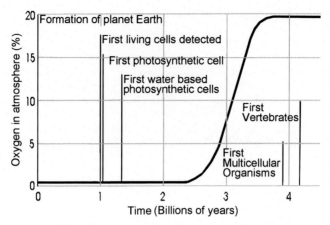

FIGURE 6.1 Evolutionary history of planet earth.

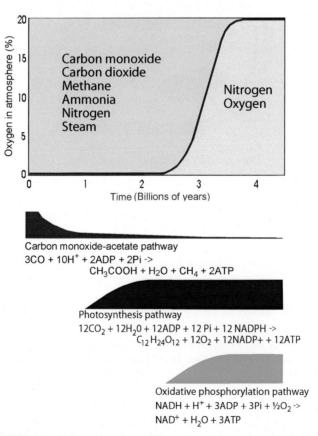

Carbon monoxide-acetate pathway
$$3CO + 10H^+ + 2ADP + 2Pi \rightarrow$$
$$CH_3COOH + H_2O + CH_4 + 2ATP$$

Photosynthesis pathway
$$12CO_2 + 12H_2O + 12ADP + 12\,Pi + 12\,NADPH \rightarrow$$
$$C_{12}H_{24}O_{12} + 12O_2 + 12NADP+ + 12ATP$$

Oxidative phosphorylation pathway
$$NADH + H^+ + 3ADP + 3Pi + \tfrac{1}{2}O_2 \rightarrow$$
$$NAD^+ + H_2O + 3ATP$$

FIGURE 6.2 Energy pathways, red, blue, and green charts show time frames.

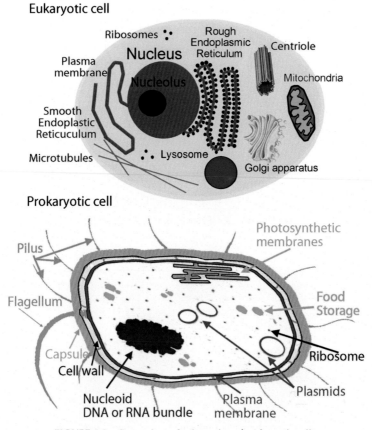

FIGURE 6.3 Comparison of eukaryotic and prokaryotic cells.

protein folding, and quality control. It is called rough because it is studded with hundreds of ribosomes. The ribosome is the site of mRNA translation or protein production in cells. Ribosomes are also found free within the cell and not just attached to the rough endoplasmic reticulum. Eukaryotes have 80S ribosomes, each consisting of a small (40S) and large (60S) subunit [4]. The smooth endoplasmic reticulum is a site of fat and steroid production in the eukaryotic cell. It is called smooth because its fat produces a greasy structure. The Golgi apparatus is an organelle found in most eukaryotic cells. The Golgi apparatus packages proteins into membrane-bound vesicles inside the cell, they are then sent to their secretory destination. It is also a site of importance in processing proteins for secretion [8].

A lysosome is a cell organelle found in most eukaryotic animal cells. Structurally and chemically, they are spherical vesicles containing hydrolytic enzymes capable of breaking down virtually all kinds of biomolecules, including proteins, nucleic acids, carbohydrates, lipids, and cellular debris.

These are a cells waste product disposal organelle that can digest unwanted materials (Fig. 6.3).

The mitochondria are a eukaryotic cell's power generator, running oxidative phosphorylation. The mitochondrion are the site of major power generating enzymatic pathways, such as the citric cycle and β-oxidation of fatty acids.

Microtubules are a component of the eukaryotic cell cytoskeleton, found throughout the cytoplasm. They provide platforms for intracellular transport and are involved in the movement of secretory vesicles and organelles. They are also involved in chromosome separation (mitosis and meiosis), and are the major constituents of mitotic spindles, used to pull apart chromosomes. Eukaryotic cell centrioles are involved in the organization of the mitotic spindle and in the completion of cytokinesis (Fig. 6.3).

Prokaryotic cells are arranged completely differently to eukaryotic cell. Prokaryotic cells do not have a nucleus as they have different organelles to eukaryotic cells. The only organelle common to eukaryotic cells are the ribosomes (Fig. 6.3). However, the ribosome of a eukaryotic cell is rather different in size and structure to that of a prokaryotic cell.

Prokaryotic cells have a nucleoid, a structure that stores the chromosomes or bundles of DNA or RNA (Fig. 6.3) [5]. The nucleoid (meaning nucleus-like) is an irregularly-shaped region containing the genetic material. In contrast to the nucleus of a eukaryotic cell, it is not surrounded by a nuclear membrane. The DNA of prokaryotic organisms is generally circular and double-stranded [5].

The prokaryotic cell capsule, protects the cell when it is engulfed by other organisms, assists in retaining moisture, and helps the cell adhere to surfaces and nutrients (Fig. 6.3). The cell wall is an outer covering that protects the cell and gives it shape. The plasma membrane surrounds the cell's cytoplasm and regulates the flow of substances in and out of the cell. The plasma membrane is selectively permeable to ions and organic molecules and controls the movement of substances in and out of cells. The plasma membrane of prokaryotic cells, like that of eukaryotic cells is a phospholipid bilayer with embedded proteins.

Pilus are hair-like structures on the surface of the prokaryotic cell that attach to other cells (Fig. 6.3). Shorter pili called fimbriae help bacteria attach to surfaces. The proteins that gather light for plant photosynthesis are embedded within cell membranes in a site called the photosynthetic membrane (or the thylakoid membrane). These proteins form the light harvesting antenna that provides energy to a number of vital photosynthetic processes such as water oxidation and oxygen evolution, the pumping of protons across the thylakoid membranes coupled with the electron transport chain of the photosystems and cytochrome b6f complex, and ATP synthesis by ATP synthase utilizing the generated proton gradient.

The plasmid is a small DNA molecule within a chamber that is physically separated from chromosomal DNA and can replicate independently [6]. They are most commonly found as small, circular, double-stranded DNA molecules. In nature, plasmids often carry genes that may benefit the survival of the organism. While the chromosomes are big and contain all the essential information for living, plasmids usually are very small and contain additional information [6].

The ribosomes in prokaryotic cells are somewhat different from those in eukaryotic cells (Fig. 6.3), although they are still the site of RNA translation or protein synthesis [4,7]. Prokaryotes have 70S ribosomes, each consisting of a small (30S) and a large (50S) subunit compared to eukaryotes which have 80S ribosomes, each consisting of a small (40S) and a large (60S) subunit [4,7]. The flagellum is a lash-like appendage that protrudes from the cell body of prokaryote cells. The word flagellum in Latin means whip. The primary role of the flagellum is locomotion but it also often functions as a sensory organelle, being sensitive to chemicals and temperatures outside the cell.

MULTICELLULAR ORGANISMS

The appearance of multicellular organisms about 1.2 millions years ago may be the third most important step in evolutionary history. Multicellular organisms are organisms that consist of multiple cells, in contrast to unicellular organisms. All animals, land plants, and filamentous fungi are multicellular, as are many algae. In contrast, a few organisms are part unicellular and part multicellular, like Dictyostelium.

There are multiple mechanisms by which multicellularity may have evolved [9,10]. One hypothesis is that a group of cells aggregated into a slug-like mass called a grex, which moved as a multicellular unit. This is essentially what slime molds do. Another hypothesis is that a cell underwent nucleus division, thereby becoming a syncytium. A membrane would then form around each nucleus and the cellular space and organelles occupied in the space, thereby resulting in a group of connected cells in one organism. A third hypothesis is that as a unicellular organism divided, the daughter cells failed to separate, resulting in a conglomeration of identical cells in one organism, which could later develop specialized tissues. This is what plants and animal embryos do [9,10].

REFERENCES

[1] Bryant DA, Frigaard NU. Prokaryotic photosynthesis and phototrophy illuminated. Trends Microbiol 2006;14:488–96.
[2] Olson JM. Photosynthesis in the Archean era. Photosyn Res 2006;88:109–17.
[3] Harris H. The birth of the cell. New Haven: Yale University Press; 1999.
[4] Palade GE. A small particulate component of the cytoplasm. J Biophys Biochem Cytol 1955;1:59–68.
[5] Thanbichler M, Wang S, Shapiro L. The bacterial nucleoid: a highly organized and dynamic structure. J Cell Biochem 2005;96:506–21.

[6] Lederberg J. Cell genetics and hereditary symbiosis. Physiol Rev 1952;32:403–30.

[7] Benne R, Sloof P. Evolution of the mitochondrial protein synthetic machinery. Bio Systems 1987;21:51–68.

[8] Difference between 70S ribosomes and 80S ribosomes, RNA, micromolecules. www.microbiologyprocedure.com.

[9] Shafiee A, McCune M, Kosztin I, Forgacs G. Shape evolution of multicellular systems; application to tissue engineering. Biophys J 2014;106:6018A.

[10] Grosberg RK, Strathmann RR. The evolution of multicellularity: a minor major transition? Annu Rev Ecol Evol Syst 2007;38:621–54.

Chapter 7

What Is Life?

What is life? This is a very big question. In the case of a very simple beginning to life, as described in Chapters 2 and 5, it is a fatty acid vesicle full of random simple chemical reactions. This is just a very simplistic chemical chamber that is barely life; it is a chemical life. If one considers human cells, by comparison, this is something extremely different, where incredible genetically controlled precise and complex reactions are occurring in a multifaceted incredible human being. This is superlife, or life at is greatest. As illustrated in Fig. 7.1, these are the extremes, and there are at least five levels of life that lie in-between.

There is Chemical Life as represented by that simplistic first life chemical chamber (Fig. 7.1). There is prokaryotic life, prokaryotic life forms during the toxic carbon monoxide atmosphere, such as bacteria relying on photosynthesis. I know that I kill flies, to my mind they are not like human life [1]. Based on this figure flies are still an advanced life form, a eukaryotic life with a brain. Would I mind killing millions of prokaryotic pre-ox life forms. My mind does not consider them as life and would wipe them away by the millions using anti-biotic, or using bleach.

There is eukaryotic life, there are protozoans, slime molds and some forms of algae [2], and fungi [2]. Life has so many life-forms, it is even trouble considering anything that does not have legs and head as being a living being [3,4]. Then there is eukaryotic life with a brain (eukaryotic brain), considered as beings. Once again I think about flies and swatting them. It is in the next life category, eukaryotic life with a highly differentiated brain (eukaryotic brain differentiated) that I start to recognize life, like dogs, cats, and monkeys differentiated life-form. This is where organism can be visually recognized [4]. Where life starts becoming valuable. Then there is eukaryotic life with advanced differentiated brain, such as humans, where life is most valuable. Fig. 7.1 tries to hypothetically correlate six life classes and meaning of life.

Let me start by asking, what is life?

From **Dictionary.com** the definition is "the condition that distinguishes organisms from inorganic objects and dead organisms, being manifested by growth through metabolism, reproduction, and the power of adaptation to environment through changes originating internally."

From **Merriam-Webster.com** the definition is "the ability to grow, change, etc., that separates plants and animals from things like water or rocks."

Biology of Life. http://dx.doi.org/10.1016/B978-0-12-809685-7.00007-1

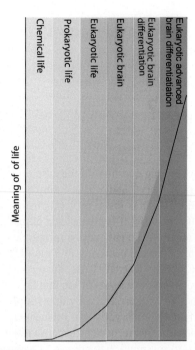

FIGURE 7.1 The hypothetical stages of life [2–4].

From **OxfordDictionaries.com** the definition is "The condition that distinguishes animals and plants from inorganic matter, including the capacity for growth, reproduction, functional activity, and continual change preceding death: the origins of life."

From **Wikipedia.com** "Life is a characteristic distinguishing physical entities having biological processes (such as signaling and self-sustaining processes) from those that do not."

Putting these four definitions together "life is the condition that separates organisms or physical entities that grow and reproduce from inorganic matter like water or rocks."

NASA is sending automated space vessels to Mars, Venus, and to Jupiter's moon Europa to determine if there is any evidence that life once existed. What substance can they search for. Can they search for DNA, RNA, Nucleotides, amino acids, adenosine triphosphate, or other molecules. All of these are indicators of life, but nothing more. The presence of any of these molecules in an empty bilayer cell does not meet our common definition of life. How about detection of a pathway intermediate like Acetyl CoA. It is suggestive of life but does not surely demonstrate life, finding acetyl CoA in an empty cell means nothing.

NASA currently has such a mission on the way to Mars. For them, detection of water means that at some time life might have existed, as does the demonstration of carbon compounds, such as organic molecules or carbon dioxide. This is very much a first stage detection of life.

Human conception of life is a creature with a head, brain, and legs. This limits life conception to just the three final categories of the life description (Fig. 7.1). It limits conception to eukaryotic life with a brain, a highly differentiated brain, or an advanced differentiated brain. I watched a Discovery Channel program "Naked and Afraid." Several times, individuals stated that they love life and would not under any circumstances kill any organism. What about that footstep they just made? They may have killed hundreds of organisms. What about that antibiotic they just took? It may have killed millions of organisms. The person's whole concept of life is limited, as stated, to the three final categories of the life description; they will not kill any birds, lizards, snakes, or more advanced life-forms.

Does the first form of life described in Chapters 2 and 5 truly meet the four definitions of life, or combined definition of life (discussed previously). It is a fatty acid vesicle containing RNA, nucleotides, and amino acids. Yes, the RNA as ribozymes is catalyzing multiple enzymatic reactions; the vesicle is effectively growing by taking in more amino acids and more nucleotides from the environment; the vesicle is reproducing by bursting and reforming as multiple vesicles. However, is it truly alive? By definition, it is growing and reproducing and meeting all juristic qualifications. But ethically and practically it is just a chamber in which chemical reactions are ongoing. You decide. It is not like complex animal life, it is just a primitive form of life. What is the difference? Animal life is just more complex, more highly controlled, and more intricately regulated.

REFERENCES

[1] Kellert SR. The value of life: biological diversity and human society. Washington, DC: Island Press; 1997.
[2] Giezen M. Mitochondria and the rise of eukaryotes. BioScience 2011;61:594–601.
[3] Yamaguchi M, Mori Y, Kozuka Y, Okada H, Uematsu K, Tame A, et al. Prokaryote or eukaryote? A unique microorganism from the deep sea. Microscopy 2012;0:1–9.
[4] Weikart R. Does Darwinism devalue human life? Human Life Rev 2004;30:29–37.

Section II

DNA and RNA

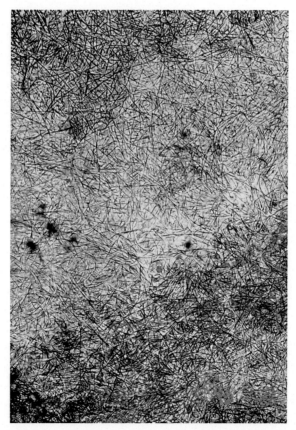

Picture taken by Laurence Cole. Glass window in the old Santa Fe Railroad workshop.

Section II

DNA and RNA

Chapter 8

The Evolution of DNA and the Genetic Code

EVOLUTION OF DNA

DNA can be thought of as a modified form of RNA, since the normal ribose sugar in RNA is reduced to deoxyribose in DNA. Similarly, the simple base uracil in RNA is methylated into thymidine in DNA. Today, the DNA precursors (the four deoxyribonucleotides, dNTPs) are produced by reduction of ribonucleotides di- or triphosphate by ribonucleotide reductases. The in vitro manufacture of DNA building blocks from RNA precursors is a major argument in favor of RNA preceding DNA in evolution. The direct prebiotic origin of it is theoretically plausible but highly unlikely, considering that evolution works like a tinkerer, not an engineer [1–3].

The first step in the appearance of DNA is most likely the formation of U-DNA or DNA containing uracil, since ribonucleotide reductases produce deoxyuridine triphosphate (dUTP) from uridine triphosphate and not deoxythymidine triphosphate (dTTP) from thymidine triphosphate, which does not exist in the cell. A few modern viruses have a U-DNA genome with uracil [4], possibly reflecting this first transition step between the RNA and DNA. The choice of the letter T occurred probably in a second step, dTTP being produced in modern cells by the modification of deoxyuridine monophosphate (dUMP) into deoxythymidine triphosphate (dTMP) by thymidylate synthases, followed by phosphorylation [5]. Fascinatingly, the same kinase can phosphorylate both dUMP and dTMP [5]. In today's cells, dUMP is produced from dUTP by deoxyuridine triphosphatases, or from deoxycytidine monophosphate (dCMP) by dCMP deaminases [5]. This is another sign that thymidine (T)-DNA originated after U-DNA. In early U-DNA cells, dUMP might have also been produced by degradation of deoxyuridine-DNA.

The evolution of DNA also required the appearance of enzymes able to incorporate dNTPs, firstly using RNA templates (reverse transcriptases) and later on using DNA templates (DNA polymerases). In all living organisms, cells and viruses, these enzymes work in the 5' to 3' direction. This directionality is dictated by the cellular metabolism that produces only dNTP 5' triphosphates and no 3' triphosphates. Both purine and pyrimidine biosyntheses are built upon ribose 5 monophosphate as a common precursor.

DNA synthesis is therefore a relic of the RNA world of metabolism. Modern DNA polymerases of the A and B families, reverse transcriptases, cellular RNA polymerases, and viral replicative RNA polymerases are structurally related and

Biology of Life. http://dx.doi.org/10.1016/B978-0-12-809685-7.00008-3

51

thus probably homologous [6]. This indicates that reverse transcriptases and DNA polymerases of the A and B families originated from an ancestral RNA polymerase that also has descendants among viral-like RNA replicases. However, there are several other DNA polymerase families whose origin is obscure.

If DNA appeared in the RNA world, it is possible to imagine that formation of the four dNTPs from the four rNTPs was initially performed by ribozymes. Most scientists, who consider that the reduction of ribose cannot be accomplished by an RNA enzyme, now reject this hypothesis [3,7–13]. The deletion of the 2′ oxygen in the ribose, indeed involves a complex chemistry for reduction that requires the formation of stable radicals in ribonucleotide reductases. Such radicals would have destroyed the RNA backbone of a ribozyme by attacking the labile phosphodiester bond of RNA. Accordingly, DNA could have only originated after the invention of modern complex proteins, in an already elaborated protein–DNA world. This suggests that RNA polymerases were indeed available at that time to evolve into DNA polymerases.

The main question is, why was DNA selected to replace RNA? The basic explanation is that DNA replaced RNA as genetic material because it is more stable and can be repaired more faithfully [1]. Clearly, removal of the 2′ oxygen of the ribose in DNA stabilized the molecule, since this reactive oxygen can attack the phosphodiester bond (explaining why RNA is so prone to strand breakage). Furthermore, the replacement of uracil by thymine has made it possible to correct the deleterious effect of spontaneous cytosine deamination, since a misplaced uracil cannot be recognized in RNA, whereas it can be pinpointed as an alien base in DNA and efficiently removed by repair systems. Changing RNA to DNA as genetic material opened the way to the formation of large genomes, a prerequisite for the evolution of modern cells.

In summary, first came RNA as the genetic material. RNA uses Adenine–Uracil and Cytosine–Guanine bases. Then RNA underwent a simple reduction losing a 2′ oxygen to make DNA. Early DNA was a U-DNA having Adenine–Uracil and Cytosine–Guanine bases; this is found on some viruses. Then in the final evolutionary step, Uracil was converted to Thymidine by thymidylate synthases, making the DNA bases, Adenine–Thymidine and Cytosine–Guanine.

EVOLUTION OF THE GENETIC CODE

The often mentioned fact that humans and chimpanzees are 99.9% identical in their DNA is hard to accept for some people, who can't comprehend how we could share so much of our basic genetic endowment even with the most human-like ape. Yet this genetic similarity is very real, and it dramatically shows how parsimonious natural selection can be; it reuses genes and structures that have worked well in the past.

It was also mind-boggling when, in 1987, British researchers demonstrated that a human gene could be inserted into the cells of a lowly yeast, and it functioned perfectly well. In this landmark experiment, researchers Paul Nurse and Melanie Lee showed that the gene in question, one that controlled the division of cells, was extremely similar despite the fact that yeast and the distant ancestors of humans diverged about 1 billion years ago.

The Human Genome Project is revealing many dramatic examples of how genes have been "conserved" throughout evolution. That is, genes that perform certain functions in lower animals have been maintained even in the human DNA script, although sometimes the genes have been modified for more complex functions.

This thread of genetic similarity connects us and the roughly 10 million other species in the modern world to the entire history of life, back to a single common ancestor more than 3.5 billion years ago (Table 8.1). The evolutionary view of a single and very ancient origin of life is supported at the deepest level imaginable; the very nature of the DNA code in which the instructions of genes and chromosomes are written. In

TABLE 8.1 The DNA Genetic Code

First base codon		Second base codon								Third base codon
		T		**C**		**A**		**G**		
	T	TTT	Phe	TCT	Ser	TAT	Tyr	TGT	Cys	**T**
		TTC	Phe	TCC	Ser	TAC	Tyr	TGC	Cys	**C**
		TTA	Leu	TCA	Ser	TAA	STOP	TGA	STOP	**A**
		TTG	Leu	TCG	Ser	TAG	STOP	TGG	Trp	**G**
	C	CTT	Leu	CCT	Pro	CAT	His	CGT	Arg	**T**
		CTC	Leu	CCC	Pro	CAC	His	CGC	Arg	**C**
		CTA	Leu	CCA	Pro	CAA	Gln	CGA	Arg	**A**
		CTG	Leu	CCG	Pro	CAG	Gln	CGG	Arg	**G**
	A	ATT	Ile	ACT	Thr	AAT	Asn	AGT	Ser	**T**
		ATC	Ile	ACC	Thr	AAC	Asn	AGC	Ser	**C**
		ATA	Ile	ACA	Thr	AAA	Lys	AGA	Arg	**A**
		ATG	Met	ACG	Thr	AAG	Lys	AGG	Arg	**G**
	T	GTT	Val	GCT	Ala	GAT	Asp	GGT	Gly	**T**
		GTC	Val	GCC	Ala	GAC	Asp	GGC	Gly	**C**
		GTA	Val	GCA	Ala	GAA	Glu	GGA	Gly	**A**
		GTG	Val	GCG	Ala	GAG	Glu	GGG	Gly	**G**

all living organisms, the instructions for reproducing and operating the individual is encoded in a chemical language with four letters, A, C, T, and G, the initials of four chemicals. Combinations of three of these letters specify each of the amino acids that the cell uses in building proteins (Table 8.1).

Biologically and chemically, there is no reason why this particular genetic code, rather than any of millions or billions of others, should exist, scientists assert (Table 8.1). Yet every species on Earth carries a genetic code that is, for all intents and purposes, identical and universal. The only scientific explanation for this situation is that the genetic code was the result of a single historic accident [14,15]. There is some logic to the genetic code, both acidic amino acids, Glu and Asp are GA_ amino acids, both basic amino acid, Lys and Arg are A_A, and A_G amino acids. Both Tyr and its aromatic variant Phe are T_T and T_C amino acids. This code was the one carried by the single ancestor of life and all of its descendants, including us [14,15].

REFERENCES

[1] Lazcano A, Guerrero R, Margulis L, et al. The evolutionary transition from RNA to DNA in early cells. J Mol Evol 1988;27:283–90.

[2] Jacob F. Evolution and tinkering. Science 1997;196:1161–6.

[3] Poole AM, Logan DT, Sjöberg B-M. The evolution of ribonucleotide reductase: much ado about oxygen. J Mol Evol 2002;55:180–96.

[4] Takahashi I, Marmur J. Replacement of thymidylic acid by deoxyuridylic acid in the deoxyribonucleic acid of a transducing phage for Bacillus subtilis. Nature 1963;197:794–5.

[5] Kornberg A, Baker T. DNA replication. New York: Freeman and Company; 1992.

[6] Ahlquist P. RNA-dependent RNA polymerases, viruses, and RNA silencing. Science 2002;296:1270–3.

[7] Freeland SJ, Knight R, Landweber LF. Do proteins predate DNA? Science 1999;286:690–2.

[8] Poole A, Penny D, Sjöberg B-M. Methyl-RNA: evolutionary bridge between RNA and DNA? Chem Biol 2000;7:207–16.

[9] Poole A, Penny D, Sjöberg B-M. Confounded cytosine! tinkering and the evolution of DNA. Nature Rev Mol Cell Biol 2001;2:147–51.

[10] Stubbe JA. Ribonucleotide reductases: the link between an RNA and a DNA world? Current Opin Structural Biol 2000;10:731–73.

[11] Eklund H, Uhin U, Farnegardh M, et al. Structure and function of the radical enzyme ribonucleotide reductase. Prog Biophys Mol Biol 2001;77:177–268.

[12] Fontecave M, Mulliez E, Logan DT. Deoxyribonucleotide synthesis in anaerobic microorganisms: the class III ribonucleotide reductase. Prog Nucleic Acid Res Mol Biol 2002;72:95–128.

[13] Myllykallio H, Lipowski G, Leduc D, et al. An alternative flavin-dependent mechanism for thymidylate synthesis. Science 2002;297:105–7.

[14] Koonin EV, Novozhilov AS. Origin and evolution of the genetic code: the universal enigma. IUBMB Life 2009;61:99–111.

[15] Davis BK. Evolution of the genetic code. Prog Biophys Mol Biol 1999;72:157–243.

FURTHER READING

[1] Murzin AG. DNA building blocks reinvented. Science 2002;297:61–2.

Chapter 9

DNA Biology: DNA Replication, Transcription, and Translation

Deoxyribonucleic acid (DNA) codes all the genetic instructions used in the function of all known living organisms. DNA is made up of the nucleotides, guanine, adenine, thymine, or cytosine (Fig. 9.1) attached to a sugar called deoxyribose and phosphate groups.

It is important not to confuse terms. Guanine, adenine, thymine, and cytosine are the four nitrogenous nucleobases of DNA. Guanine, adenine, uracil, and cytosine, in contrast, are the four nitrogenous nucleobases of ribonucleic acid (RNA). When a base is attached to a carrier sugar like deoxyribose in DNA and ribose in RNA it becomes a nucleoside and takes the name deoxy-adenosine, deoxy-guanosine, deoxy-thymidine and deoxy-cytidine in DNA or adenosine, guanosine, cytidine and uridine (uridine replaces thymidine in RNA) as in RNA. When phosphate groups are added, the nucleoside becomes a nucleotide and becomes deoxy-adenosine monophosphate (A), deoxy-guanosine monophosphate (G), deoxy-thymidine monophosphate (T) and deoxy-cytidine monophosphate (C) as in DNA and adenosine monophosphate (A), guanosine monophosphate (G), uridine monophosphate (U) and cytidine monophosphate (C) as in RNA. It is strictly the nucleotide that comprises DNA and RNA (Fig. 9.1).

The DNA and RNA nucleotides are connected to one another in a chain in DNA and RNA by covalent bonds between the sugar of one nucleotide and the phosphate of the next, resulting in an alternating sugar–phosphate backbone. In DNA, according to base pairing rules, adenine nucleotides always base pair with thymidine nucleotides, and guanine nucleotides always base pair with cytosine nucleotides. The base pairing is caused by hydrogen bonds between the nucleotides (Fig. 9.1). If one linked strand of DNA is comprised of -G-A-T-C-G-A-T-C-, it base pairs with the corresponding second strand comprised of -C-T-A-G-C-T-A-G-. The result is that the two strands or base pairs are attached to each other in DNA. The two strands fold around each other forming the classical DNA double helix structure (Fig. 9.2).

The base paired strands of DNA store the same biological information as each other, -G-A-T-C-G-A-T-C- storing the "sense" and -C-T-A-G-C-T-A-G- storing the "antisense" sequence of the genetic code. Biological information is

Biology of Life. http://dx.doi.org/10.1016/B978-0-12-809685-7.00009-5

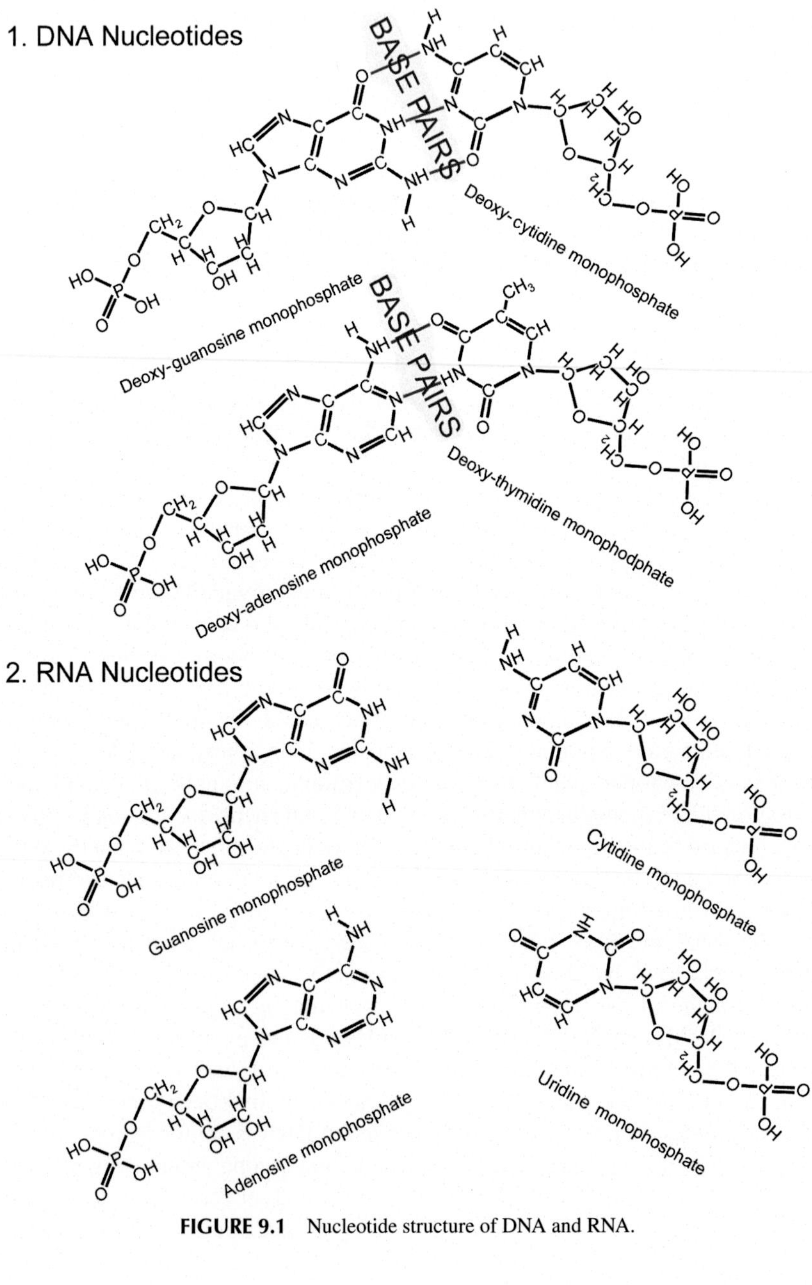

1. DNA Nucleotides

2. RNA Nucleotides

FIGURE 9.1   Nucleotide structure of DNA and RNA.

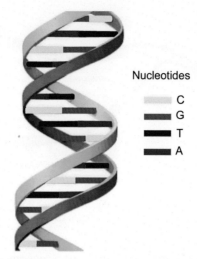

Nucleotides

▬ C
▬ G
▬ T
▬ A

FIGURE 9.2 Double stranded base-paired DNA.

effectively replicated as the two strands are separated. The DNA sequence contains the genetic code and is translated into amino acids. A significant portion of the DNA sequence, more than 98% for humans, is noncoding, meaning that these sections do not serve a function for encoding amino acids.

Within cells, DNA is arranged in long structures called chromosomes which contain 100–220 million base pairs (A, G, T, and C sequence nucleotides). During cell division, these chromosomes are duplicated providing each cell with its own complete set of chromosomes. Humans store most of their DNA inside the cell nucleus and some of their DNA in organelles, such as the mitochondria. Within the chromosomes, a protein called chromatin compacts and folds around the DNA. The folded proteins that surround and expose DNA are called histones, and the DNA–protein complex are called nucleosomes. These compact structures guide the interactions between DNA and other proteins, controlling which parts of the DNA are transcribed and translated to make needed proteins and enzymes in a specific cell.

In the double helix structure, the two stands of DNA run antiparallel. The asymmetric ends of DNA strands are called the 5′ (five prime) and 3′ (three prime) ends, with the 5′ end having a terminal phosphate group and the 3′ end a terminal hydroxyl group. One strand runs 3′ to 5′ and the base paired strand runs in the opposite direction (Fig. 9.3).

The DNA codes for specific amino acids using a group of three nucleotides or codons. The DNA genetic code is listed in Table 8.1, and shows the codon code for all 20 common amino acids. Three codons TAA, TAG, and TGA are termination codons, coding for the sequence to stop translating.. A copy of the DNA sequence of a gene coding for a specific protein or enzyme is made, this is an RNA copy called messenger RNA (mRNA). The copying process is called

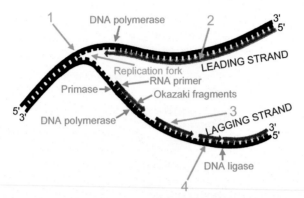

FIGURE 9.3 DNA replication. The black cog-like structure is a double helix DNA (left side) and the annealed sense and antisense strands (center and right side).

transcription. The mRNA transcript leaves the nucleus and goes to the ribosome where it is read and translated through a transfer RNA (tRNA) into a tRNA conveyed amino acid sequence. This process is called translation.

A gene sequence contains coding sequences (for making proteins and enzymes) called "exons", and noncoding sequences called "introns." An intron is any nucleotide sequence within a gene that is removed by RNA splicing while the final protein-coding mRNA is being generated. Sequences that are joined together to form the final mRNA sequence after RNA splicing are exons. When proteins are to be made from an mRNA strand, splicing takes place following transcription and preceding formal translation in the ribosome. RNA differs somewhat from DNA in not having the nucleoside thymidine. In place, RNA uses the nucleoside uridine (uracil-ribose). The four distinct nucleotides comprising RNA, in which nucleosides are formed with the sugar ribose rather that deoxyribose, are adenosine monophosphate (A), guanosine monophosphate (G), cytidine monophosphate (C), and uridine (uracil-ribose) monophosphate (U). Uridine nucleotides, like thymidine nucleotides, base pairs with adenosine nucleotides. The mRNA genetic code is essentially identical to the DNA genetic code except that U replaces T. mRNA does not base pair and double strand like DNA but remains single stranded.

Four distinct DNA and RNA pathways occur, "replication," in which the DNA sequence is duplicated as needed in cell division, "transcription," in which mRNA make a copy of a gene sequence, mRNA splicing in which exons are dissected, and "translation" in which the mRNA sequence is converted into the coded amino acids in ribosomes.

REPLICATION

Replication is the protocol whereby DNA, genes, or chromosomes duplicate themselves when cells divide. The process of DNA replication is illustrated in Fig. 9.3 and described by steps and numbers. In the first step, an enzyme called a helicase breaks apart the base paired two DNA strands. The function of helicase is to unpackage the DNA genes. They are motor proteins that move directionally along the DNA double helix separating the two annealed nucleic acid strands using energy derived from adenosine triphosphate hydrolysis (Fig. 9.3, Arrow 1).

The DNA separates into a leading strand of DNA which is replicated in the same direction as the replication fork, and a lagging strand of DNA which is replicated in the opposite direction of the replication fork. Because of its orientation, replication of the lagging strand is more complicated than that of the leading strand.

The replication fork is a structure that forms during DNA replication. The replication fork is created by helicases which cleave the hydrogen bonds holding the two DNA strands together. These two strands serve as the template which will be created as DNA polymerase matches complementary nucleotides to the separated strand templates. DNA is always synthesized in the 5′ to 3′ direction. Since the lagging strand templates are oriented in opposite directions of the replication fork, a major issue is how to achieve synthesis of new lagging strand DNA in a direction of synthesis opposite to the direction of the growing replication fork.

On the leading strand, a DNA polymerase "reads" the leading strand template and adds free complementary nucleotides to the nascent leading strand on a long continuous basis (Fig. 9.3, Arrow 2). The DNA polymerase on the leading strand in humans is DNA polymerase epsilon. On the lagging strand a DNA polymerase delta uses the free complementary nucleotides to synthesize DNA in short separated segments. A primase reads the template DNA and initiates synthesis of a short complementary RNA primer (Fig. 9.3, Arrow 3). The RNA primer is replaced by DNA. The DNA polymerase delta extends the primed segments, forming Okazaki fragments or short fragments of DNA. The fragments of DNA are joined together by a DNA linking enzyme, DNA ligase (Fig. 9.3, Arrow 4). This way, two new double helix strands of DNA are formed and a double helix structure results.

TRANSCRIPTION

Transcription is the process whereby proteins and enzymes are made exactly as coded on the sense DNA sequence. A segment of DNA is unwound or hydrogen bonds are broken exposing a segment of the two DNA strands. An mRNA copy is made or complementary copy of only the antisense DNA strand, adenosine monophosphate bases matching up with deoxy-T bases, uridine (not thymidine since it is RNA) monophosphate bases matching up with deoxy-A bases, guanosine monophosphate bases matching with deoxy-C bases, and cytidine monophosphate with deoxy-G bases. The antisense mRNA dupe is effectively an mRNA copy of the DNA sense sequence except that uridine replaces thymidine nucleotides. The mRNA leaves the nucleus through nuclear pores and

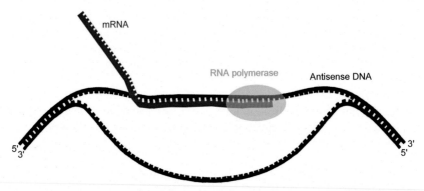

FIGURE 9.4 DNA transcription.

goes to ribosomes where the protein in constructed by tRNA molecules carrying specific amino acids.

An activator protein specifically binds the gene selecting it for transcription. Proteins called transcription factors unwind the segment of DNA strands and an RNA polymerase transcribes the single antisense strand of DNA to make the mRNA (Fig. 9.4). The strand that serves as the template for transcription is always the antisense strand. The strand that is not transcribed is called the sense strand. RNA polymerase moves along the DNA copying the base sequence until it reaches a terminator sequence (see Table 8.1). At that point, RNA polymerase releases the mRNA polymer and it detaches from the DNA (Fig. 9.4). The 3′ terminus of the new mRNA gets a long poly-A sequence tail. As many as 100 adenosine-monophosphate residues are lined up at this terminus to identify the mRNA.

mRNA SPLICING

While the mRNA molecule copies the gene on the chromosome in the nucleus, regions to be translated as amino acids in a protein (exons) need to be separated from unread regions (introns). Splicing is carried out in a series of reactions which are catalyzed by the spliceosome, a complex of RNA and protein. As part of the RNA processing pathway, introns are removed by RNA splicing either shortly after or concurrent with transcription.

Within the sequences to be removed are consensus sequences that identify the introns. Introns invariably start at 5′ terminus with a GU identifier, and invariably end at the 3′ terminus with an AG identifier. Sequences identifying the introns include M-A-G-[break]-G-U-R-A-G-U, C-U-R-[A]-Y and Y-rich-N-C-A-G. These sequences fold and instruct the spliceosome exactly where to cut and rejoin the mRNA (Fig. 9.5). The final mRNA contains only exons or only the sequence to be translated in the ribosome.

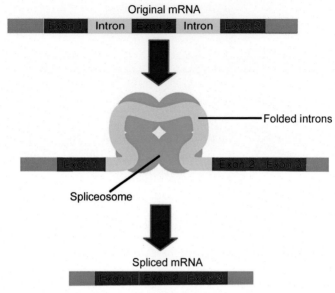

FIGURE 9.5 Spliceosome splicing of mRNA.

TRANSLATION OF mRNA

The mRNA leaves the nucleus through nuclear pores and travels to a ribosome. Here translation or conversion of RNA into an amino acid sequence occurs. The process of translation is illustrated in Fig. 9.6.

tRNA is an adaptor molecule composed of RNA, which is 73 to 94 ribonucleotides in length. tRNA is the link between the nucleotide sequence of nucleic acids and the amino acid sequence of proteins. This link is achieved by tRNA carrying an amino acid and being coded by a three nucleotide antisense sequence. When mRNA codes for the amino acid Cys, for instance, the mRNA coding sequence is UGC (Table 8.1). A tRNA carries the amino acid Cys. It contains the antisense codon, ACG. This ACG antisense sequence binds the coding mRNA, placing the amino acid Cys in the protein (Fig. 9.6).

The covalent attachment of the specific amino acid to a tRNA 3′ end is catalyzed by one of a group of enzymes called aminoacyl-tRNA synthetases. During protein synthesis, tRNAs with attached amino acids are delivered to the ribosome by proteins called elongation factors, eEF-1, which aid in decoding the mRNA codon sequence. If the tRNA's anticodon matches the mRNA codon, another tRNA already bound to the ribosome transfers the growing polypeptide chain from its 3′ end to the amino acid attached to the 3′ end of the newly delivered tRNA, a reaction catalyzed in ribosomes (Fig. 9.6).

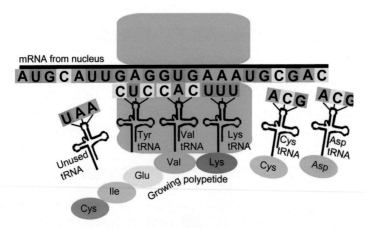

FIGURE 9.6   Translation of mRNA.

GENES AND CHROMOSOMES

Human cells contain 23 pairs of chromosomes, 23 from the individual's father and 23 from the mother. Sexually, the 23 pairs include an X and a Y chromosome in males and two X chromosome in females. Genes code for proteins including, enzymes, glycoproteins (proteins made from amino acids, which then receive carbohydrate side chains in the rough endoplasmic reticulum and Golgi apparatus of cells), structural proteins, hormones, muscular proteins, and general cellular proteins.

The 23 pairs of chromosomes each contain the coding for hundreds of genes linked head to tail. Approximately 25,000 genes fill the human 23 chromosomes in total. The DNA double helix is packaged by proteins into a condensed structure called chromatin. This allows the very long DNA molecules to fit into the cell nucleus. Chromosomes must be replicated, divided, and passed successfully to their daughter cells so as to ensure the genetic diversity and survival of their progeny. Humans possess multiple large linear chromosomes contained in the cell's nucleus. Each chromosome has one centromere, with one or two arms projecting from the centromere.

FURTHER READING

[1] Allison LA. Fundamental molecular biology. Blackwell Publishing; 2007.
[2] Berg JM, Tymoczko JL, Stryer L, Clarke ND. Imperfect DNA replication results in mutations. In: Chapter 27: DNA Replication, Recombination, and Repair. Biochemistry. W.H. Freeman and Company; 2002.
[3] Hausner W, Thomm MM. Events during initiation of archaeal transcription: open complex formation and DNA-protein interactions. J Bacteriol 2001;183:3025–31.
[4] Campbell N. Biology. 4th ed. The Benjamin Cummings Publish Co; 1996.

Section III

Energetics

Picture taken by Laurence Cole. Chandelier at the Ice Hotel, Arctic Circle, Sweden, 2010.

Chapter 10

Adenosine Triphosphate Energetics

Adenosine triphosphate (ATP) is at the root of all organisms energetics (Fig. 10.1). ATP provides the energetics for all muscle movements, heart beats, nerve signals and chemical reactions inside the body. It is estimated that the human body uses roughly 2×10^{26} transient molecules of ATP or more than the bodies weight; 160 kg of ATP in a day [1–4]. ATP stores energy in a high energy phosphate bond, the third phosphate bond (Fig. 10.1). The cutting of one phosphate bond, $ATP + H_2O \rightarrow ADP + Pi$ liberates about 30.6 kJ/mole (Fig. 10.2) [1–4].

ATP is used by eukaryotes, plants, and bacteria to power all chemical reactions and movements.

Energy is commonly freed from the ATP molecules to do work in the cell by a reaction that removes one phosphate group, leaving adenosine diphosphate (ADP) (Fig. 10.2). The ADP is recycled in the mitochondria where it is recharged again into ATP. The immense amount of activity that occurs inside each of the human body's approximately 100 trillion plus cells is exposed by the fact that at any instant each cell contains about one billion ATP molecules [5]. This is sufficient for a cell's needs for just a few minutes and must be rapidly recycled. With 100 trillion cells, about 1023 or 1 sextillion ATP molecules normally exist in the body [5]. For each ATP molecule, the terminal phosphate is added and removed roughly 3 times each minute [6].

The human body's content of ATP at any one time is approximately 50 g, which must be continuously recycled. The source of energy for constructing ATP is food; ATP is simply the carrier and storage unit of energy. An average daily intake of 2500 calories of food translates into a massive 160 kg of ATP [6].

ATP is generated by four pathways in prokaryotes and eukaryotes, through oxidative phosphorylation in eukaryotes and some advanced plants; through photosynthetic pathways in prokaryote chloroplasts; through cytoplasmic membrane electron transport in bacteria; and seemingly, through the primitive carbon monoxide–acetate pathways in some early primitive species (Chapter 5).

In eukaryotes, oxidative phosphorylation is the power generator or means for generating ATP.

Biology of Life. http://dx.doi.org/10.1016/B978-0-12-809685-7.00010-1

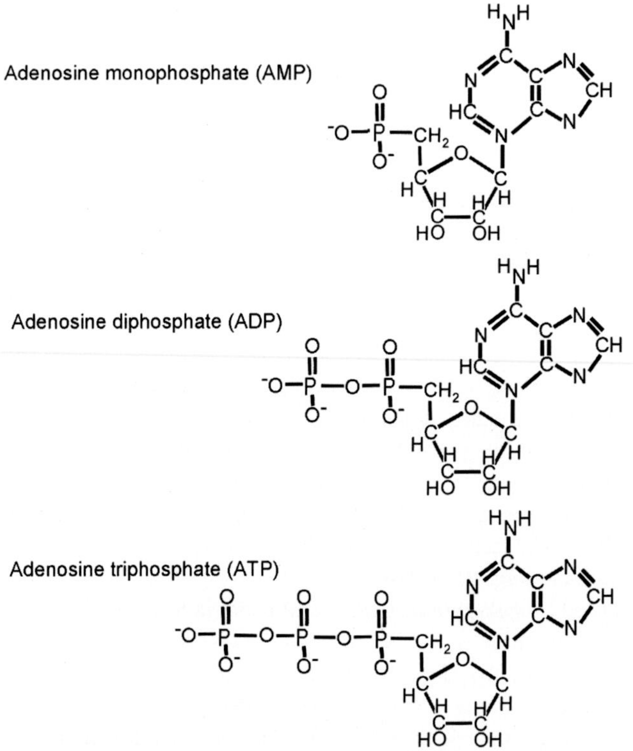

Adenosine monophosphate (AMP)

Adenosine diphosphate (ADP)

Adenosine triphosphate (ATP)

FIGURE 10.1 Structures of adenosine trisphosphate (ATP), adenosine diphosphate (ADP), and adenosine monophosphate (AMP).

| ATP | + | Water | -> | ADP | + | Pi | + | Energy |

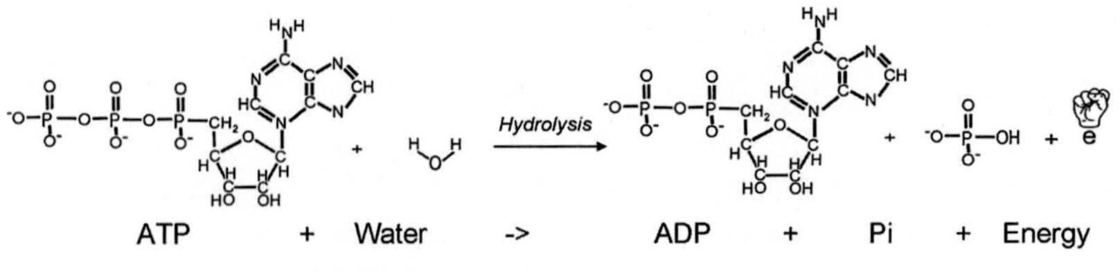

FIGURE 10.2 ATP, the energy source of life.

OXIDATIVE PHOSPHORYLATION

Oxidative phosphorylation occurs in the mitochondria of eukaryotes, including some advanced eukaryote plants. This process occurs in specially constructed chambers located in the mitochondrion's inner membranes. The mitochondrion itself functions to produce an electrical chemical gradient by accumulating hydrogen ions or protons in the space between the inner and outer mitochondrion membrane. Most of the energy comes from the enzyme chains in the

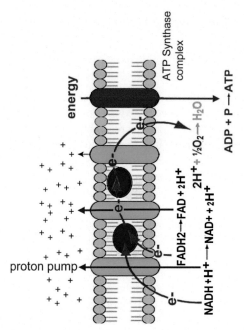

FIGURE 10.3   Oxidative phosphorylation in mitochondria.

membranous sacks on the mitochondrial walls. Most of the food energy for most organisms is produced by the electron transport chain. Cellular oxidation in the citric acid cycle causes an electron build up that is used to push proton ions outward across the inner mitochondrial membrane [7].

When the charge builds up, it provides an electrical potential causing a flow of proton ions across the inner membrane of the mitochondrion into the inner chamber (Fig. 10.3). The energy causes an enzyme to be attached to ADP which catalyzes the addition of a third phosphorus to form ATP. Eukaryote plants can also produce ATP in this manner in their mitochondria, but plants can also produce ATP by using the energy of sunlight in chloroplasts as discussed later. In the case of eukaryotic animals the energy comes from food which is converted to pyruvate and then to acetyl coenzyme A (acetyl CoA). Acetyl CoA then enters the citric acid cycle which releases energy that results in the conversion of ADP into ATP.

Chemically, oxidative phosphorylation through electron transport in the membrane of mitochondria drives the reaction of energy of nicotinamide adenine dinucleotide hydrogen (NADH) and the energy of flavin adenine dinucleotide hydrogen (FADH$_2$) along with oxygen to power ATP synthesis (Fig. 10.3). Each NADH molecule powers the generation of three ATP molecules and each FADH$_2$ molecule powers a shorter electron transport cycle or the generation of two ATP molecules (Fig. 10.3).

$$1 \text{ NADH} + \frac{1}{2} O_2 + 3 \text{ ADP} \rightarrow 1 \text{ NAD}^+ + H_2O + 3 \text{ ATP}$$

$$1 \text{ FADH}_2 + \frac{1}{2} O_2 + 2 \text{ ADP} \rightarrow 1 \text{ FAD} + \_ H_2O + 2 \text{ ATP}$$

Oxidative phosphorylation is the principal purpose of oxygen respiration and the principal use of breathed in oxygen to generate energy in the body.

The ATP synthase revolving door resembles a molecular water wheel that harnesses the flow of hydrogen ions in order to build ATP molecules. Each revolution of the wheel requires the energy of about nine hydrogen ions returning into the mitochondrial inner chamber [8]. Located on the ATP synthase are three active sites, each of which converts ADP to ATP with every turn of the wheel. Under maximum conditions, the ATP synthase wheel turns at a rate of up to 200 revolutions per second, producing 600 ATPs during that second [5].

Although ATP contains the amount of energy needed for most reactions, sometimes more energy is required. The solution is for ATP to release two phosphates instead of one, producing an adenosine monophosphate (AMP) plus two phosphates or what we call pyrophosphate. How AMP is built up into ATP again illustrates the complexity of the cell energy system.

The main energy carrier in the body is ATP. Other energized nucleotides are also utilized such as thymine, guanine, uracil, and cytosine for making RNA and DNA. The Krebs cycle charges only ADP, but the energy contained in ATP can be transferred to one of the other nucleosides by means of an enzyme called nucleoside diphosphate kinase. This enzyme transfers the triphosphate from a nucleoside triphosphate, commonly ATP, to a nucleoside diphosphate such as guanosine diphosphate to form guanosine triphosphate.

FEEDING PATHWAYS TO OXIDATIVE PHOSPHORYLATION

In eukaryotes and advanced eukaryote plant chemistry, the pathways of glycogenolysis, sugar glycolysis, fatty acid oxidation, and of amino acid break up, clearly feed into oxidative phosphorylation as energy sources.

Glucose in muscle is very limited. Glucose digestion, glycolysis, provides 2 ATP molecules and 4 NADH molecules (Fig. 10.4). Two pyruvic acid molecules generated undergo 2 rounds of the citric acid cycle generating 6 NADH and 2 $FADH_2$ molecules. The NADH molecules make 30 ATP during oxidative phosphorylation. The 2 $FADH_2$ generate 4 ATP molecules plus the 2 ATP synthesized in glycolysis, makes 36 ATP molecules per glucose molecule.

Fatty acids are a much bigger source of ATP generation. Palmitic acid (16 carbons) is a common fatty acid. Taking a fatty acid like palmitic acid through 8 rounds of β-oxidation in the mitochondrion generates 1 NADH and 1 $FADH_2$ in each round, or 8 NADH and 8 $FADH_2$ (Fig. 10.5). This generates 8 Acetyl CoA molecules. Each Acetyl CoA molecule enters the citric acid cycle, each generating 3 more NADH and 1 more $FADH_2$ molecule. As such, palmitic acid breakdown generates 24 NADH and 16 $FADH_2$ molecules. Through oxidative

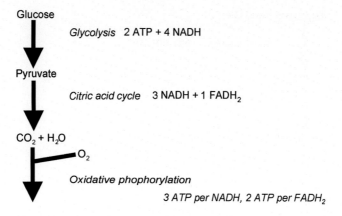

2 ATP from glycolysis
4 NADH at glycolysis generates 12 ATP
3 NADH at citric acid cycle, 2 cycles generates 18 ATP
FADH2 at citric acid cyle, 2 cycles generates 4 ATP
Total ATP generated is 36

FIGURE 10.4   Use of glucose and glycolysis as a source of energy in the citric acid cycle.

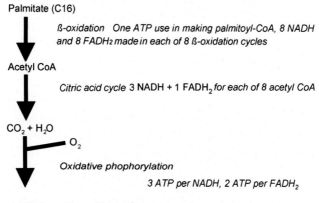

-1 ATP for making palimitoyl-CoA.
8 NADH made in ß-oxidation cycles generates 24 ATP
8 FADH2 made in ß-oxidation cycles generates 16 ATP
Citric acid cycle generates 3 NADH + 1 FADH2 per Acetyl CoA or 88 ATP
Total ATP generated is 127

FIGURE 10.5   Use of fatty acids as a source of energy, for example palmitic acid.

phosphorylation, these all generate 96 (NADH) + 32 (FADH$_2$) ATP molecules or 128 ATP molecules. Palmitic acid uses one ATP molecule in making palmitoyl-CoA. As such, the balance is 127 ATP molecules.

Glycogen is a major source of energy which is stored in muscle and the liver. Glycogen is broken down to glucose releasing its massive source of energy. Glucose is broken down by glycolysis and the citric acid cycle

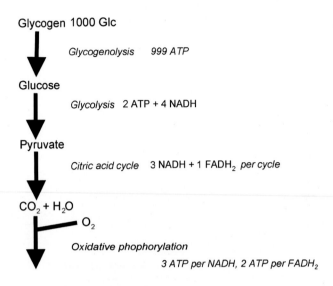

Glycogen 1000 Glc

Glycogenolysis   *999 ATP*

Glucose

Glycolysis   2 ATP + 4 NADH

Pyruvate

Citric acid cycle   3 NADH + 1 FADH$_2$ *per cycle*

$CO_2 + H_2O$

O_2

Oxidative phophorylation

           3 ATP per NADH, 2 ATP per FADH$_2$

999 ATP from glycogenolysis
2 x 1000, 2000 ATP from glycolysis
4 NADH at glycolysis generates 12,000 ATP
3 NADH at citric acid cycle, 2 cycles generates 18,000 ATP
FADH2 at citric acid cyle, 2 cycles generates 4,000 ATP
Total ATP generated is 36,999 ATP

FIGURE 10.6   Use of glycogen as a source of energy. The example glycogen is composed of 1000 glucose residues.

(Fig. 10.6). Both glycogen stored in the muscle and stored in the liver are used as a major supply of glucose. The average glycogen molecule contains β1, 4 linked glucose chains of 8–12 glucose molecules, α1, 6 branched to more chains. The entire molecule may contain as many as 30,000 glucose molecules. In the center of the glycogen molecule is a glycogenin making glycogen into a glycoprotein. A 30,000 glucose glycogen can make more than 1,000,000 ATP molecules.

Amino acids and proteins in muscles can also power muscles through an ATP pathway (Fig. 10.7). Proteins are digested with proteases and then the amino acids are broken down as illustrated in Fig. 10.7.

The significance of the body's oxidative phosphorylation system for energy production is illustrated from the finding of what energy sources drive the body's muscles. It is estimated, considering energy supplies stored in muscles and collagen stored in the liver, that a muscle may be primarily powered by a combination of glycogen in the muscle, glycogen in the liver, fatty acids in the muscle, and proteins in the muscle. With approximately 480 kcal of energy coming from glycogen in the muscle, approximately

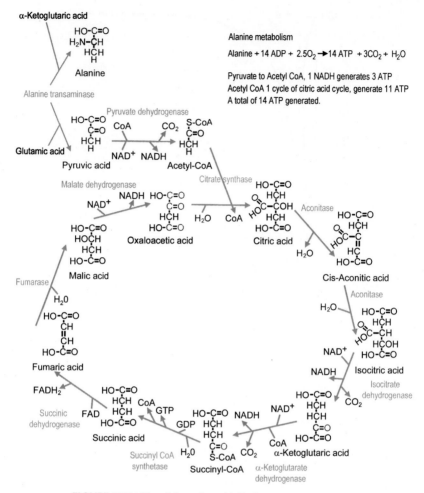

FIGURE 10.7 Use of the amino acid Alanine as an energy source.

280 kcal from glycogen in the liver, approximately 141,000 kcal coming from fatty acids, and approximately 24,000 kcal from proteins in the muscle [1–4,9].

USE OF ADENOSINE TRIPHOSPHATE ENERGY

ATP provides the energy to drive all muscles. Actin and myosin in all muscles follows the equation:

$$ATP + Actin + Myosin \rightarrow Actomyosin + Pi + ADP + Energy$$

There are seemingly three clear ATP pathways that can drive muscles, the immediate energy pathway, the nonoxidative quick energy pathway, and the

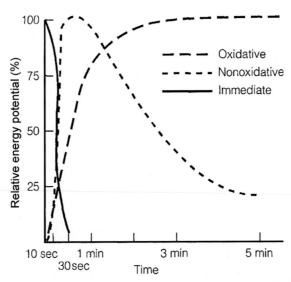

FIGURE 10.8   The three energy sources, as published by Edington and Edgerton [10].

oxidative energy pathway. The 1976 studies of Edington and Edgerton illustrate the three pathways. The immediate pathway is the energy that can cover an immediate or 10 s event, the nonoxidative pathways is the energy which can cover a 2 min event, and the oxidative pathway is the energy with can cover a greater than 2 min event. By analogy, in athletics, the immediate event might be shot put, the nonoxidative event might be the 100 m run, and the oxidative event might be the marathon run.

These three energy sources are illustrated in Fig. 10.8. The immediate energy source is the ATP stored in muscle cells, with the adenylate kinase pathways waiting to supplement the ATP supply in the immediate energy (Fig. 10.9) by combining two ADP molecules to make AMP. The creatine phosphate energy supply is also waiting in the muscle in reserve to top-off the used ATP sources (Fig. 10.9).

The nonoxidative rapidly available energy source is the ATP, and not the NADH produced by glycolysis (Fig. 10.9), coming from the degradation of glycogen that is processed to pyruvate and to lactic acid.

The oxidative energy is the massive amount of ATP formed by oxidative phosphorylation of NADH and $FADH_2$. With energy coming from glycogen, fatty acids, and amino acids (Fig. 10.9).

ADENOSINE TRIPHOSPHATE AND THE HEART

The heart needs ATP for the membrane transport systems like the Na+/K+ system ATPase, as well as for sarcomere contraction and relaxation, which involve myosin ATPase and ATP-dependent transport of calcium by the sarcoplasmic

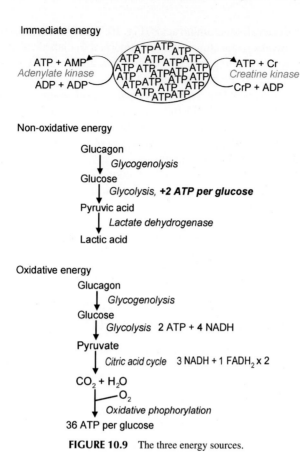

Immediate energy

ATP + AMP
Adenylate kinase
ADP + ADP

ATP + Cr
Creatine kinase
CrP + ADP

Non-oxidative energy

Glucagon
↓ *Glycogenolysis*
Glucose
↓ *Glycolysis, +2 ATP per glucose*
Pyruvic acid
↓ *Lactate dehydrogenase*
Lactic acid

Oxidative energy

Glucagon
↓ *Glycogenolysis*
Glucose
↓ *Glycolysis* 2 ATP + 4 NADH
Pyruvate
↓ *Citric acid cycle* 3 NADH + 1 $FADH_2$ x 2
$CO_2 + H_2O$
—O_2
↓ *Oxidative phophorylation*
36 ATP per glucose

FIGURE 10.9 The three energy sources.

reticulum. Therefore, increasing the physical activity of the heart by increasing heart rate and contractility increases myocardial metabolism.

Heart cellular ATP pools depend on the balance between ATP use and ATP production. The heart has an absolute requirement for aerobic oxidative production of ATP from oxidative phosphorylation to maintain needed ATP concentrations. Anaerobic capacity or nonoxidative sources are limited in the heart. Cellular ATP levels will fall if there is inadequate O_2 available to produce ATP aerobically, or if there is an increase in ATP utilization that is not coordinated by a parallel increase in ATP production.

The heart uses many substrates to oxidatively regenerate ATP. In the postabsorptive state after a meal, the heart utilizes mostly fatty acids (60–70%) and glycogen (~30%) to oxidatively generate ATP. Following a high carbohydrate meal, the heart can adapt itself to utilize carbohydrates exclusively. Lactate can be used in place of glucose, and becomes an important substrate during exercise (Lactate → pyruvate → citric acid cycle, generates 4 NADH and 1 $FADH_2$ or 14

ATP). The heart can also use amino acids (Fig. 10.7) and ketones instead of fatty acids. Ketone bodies (ie, acetoacetate) are important in diabetic acidosis [11].

During ischemia and hypoxia, the coronary circulation is not able to deliver metabolic substrates to the heart to support oxidative metabolism. Under these conditions, the heart is able to utilize glycogen as a substrate for production of ATP. However, the amount of ATP that the heart can produce by this pathway is very small compared to the amount that can be produced via oxidative metabolism. Furthermore, the heart has a limited supply of glycogen, which is rapidly depleted under severely hypoxic conditions.

The heart has a high rate of ATP production and turnover required to maintain its continuous mechanical work. Perturbations in ATP-generating processes may therefore affect contractile function directly. Characterizing cardiac metabolism in heart failure revealed several metabolic alterations called metabolic remodeling, ranging from changes in substrate use to mitochondrial dysfunction, ultimately resulting in ATP deficiency and impaired contractility [12].

PHOTOSYNTHESIS AND THE GENERATION OF OXYGEN AND ADENOSINE TRIPHOSPHATE

Chloroplasts are ATP-producing organelles found in plants. Inside is a set of thin membranes organized into flattened sacs stacked-like coins called thylakoids. The disks contain chlorophyll pigments that absorb solar energy which is the source of energy for all the plant's needs including manufacturing carbohydrates from carbon dioxide and water [13] (Fig. 10.10).

The chloroplasts first processes the solar energy into ATP (Fig. 10.11). This is then used to manufacture glucose by the Calvin Cycle and Glycolysis (Fig. 5.2).

The chloroplast also has an electron transport system for producing ATP (Fig. 10.11). The electrons are taken from water. During photosynthesis, carbon

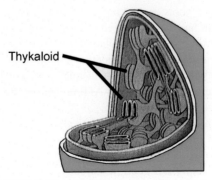

Thykaloid

FIGURE 10.10 Thykaloid in a chloroplast in a plant.

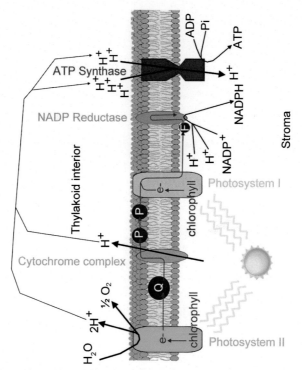

FIGURE 10.11   The light stage of photosynthesis. The abbreviation Q is plastoquinone, P is Plastocyanin, and F is ferredoxin NADP-reductase.

dioxide is reduced to glucose by the Calvin cycle and Glycolysis using energy obtained from ATP [13].

$$CO_2 + Ribulose\text{-}1, 5\text{-}Bisphosphate \rightarrow 2 \times 3\text{-}Phoshoglycerate\ \textit{Calvin cycle}$$

$$2 \times 3\text{-}Phosphoglycerate \rightarrow 2 \times Glyceraldehyde\text{-}3\text{-}Phosphate \rightarrow Glucose\ \textit{Glycolysis}$$

CYTOPLASMIC MEMBRANE ELECTRON TRANSPORT AND ADENOSINE TRIPHOSPHATE SYNTHESIS IN BACTERIA

In bacteria, the ATPase activity is located inside the cytoplasmic membrane between the hydrophobic tails of the phospholipid membrane inner and outer walls. Breakdown of sugar and other food causes the positively charged protons on the outside of the membrane to accumulate a higher concentration of protons than on the inside of the membrane. This creates a major positive charge on the outside of the membrane and a relatively negative charge on the inside.

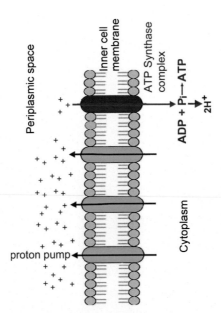

FIGURE 10.12 ATP synthesis in bacteria.

The result of this charge difference is a dissociation of H_2O molecules into H^+ and OH^- ions. The H^+ ions that are produced are transported outside of the cell and the OH^- ions remain on the inside. This results in a potential energy gradient. The force that the potential energy gradient produces is a *proton motive force* that can accomplish a variety of cell functions including converting ADP into ATP [14] (Fig. 10.12).

The proton gradient that results drives ATP synthesis by use of the ATPase complex. This allows bacteria to live in low oxygen but rich light regions. This anaerobic ATP synthesis system, unique to prokaryotes, uses a chemical compound other than oxygen as a terminal electron acceptor. The location of the ATP producing system is only one of many major contrasts that exist between bacterial cell membranes and mitochondria [14] (Fig. 10.12).

EVOLUTION OF ENERGETICS

Very clearly, all energetics or ATP generation systems are similar, whether in bacteria, eukaryotes, or plants. All three systems involve a proton gradient or electron transport as the source of energy which drives ATP synthesis. It appears that the simplest system is the bacterial membrane system, and that the more thorough photosynthesis system evolved from the bacterial system by the addition of chlorophyll and light, and that the mitochondrial oxidative phosphorylation evolved from the bacterial membrane system by the addition of the element oxygen.

ENERGETICS AND LIFE

In the first section of this book, "How life started", I described the first possible form of life. What I described as the first form of life are bundles of chemicals and chemical reactions wrapped up in a fatty acid bilayer membrane. Is this really life I ask, it is just a bundle of chemicals? It is my opinion that for something to be defined as really alive, it needs to be many times more fascinating. Like a cell that is driving an energy reaction, like oxidative phosphorylation, photosynthesis, or bacterial electron transport. This is real life. This is life that is driven by a power or energy station.

REFERENCES

[1] Brooks GA, Fahey TD, Baldwin K. Exercise physiology: human bioenergetics and its application with Powerweb Bind-in Card/Edition 4. Ch 3. The maintenance of ATP Homeostasis in energestics and human movement. Columbus, OH: McGraw Hill Education; 2004.

[2] Pikart CM, Jencks WP. Energetics of the calcium-trasporting ATPase. J Biol Chem 1984;259:1629–34.

[3] Brooks GA. Mammalian fuel utilization. During sustained exercise. Comp Biochem Physiol 1998;120:89–107.

[4] Gaessner GA, Brooks GA. Muscular efficiency during steady-rate exercise: effects of speed and work rate. J Appl Physiol 1975;38:29–43.

[5] Bergman J. ATP: The perfect energy currency for the cell. 1999. http://www.trueorigin.org/atp.php.

[6] Kornberg A. For the love of enzymes. Cambridge, MA: Harvard University Press; 1989. p. 65.

[7] Hickman CP. Integrated principles of zoology. 10th ed. New York: McGraw Hill; 1997.

[8] Goodsell DS. Our molecular nature. New York: Springer-Verlag; 1996. p. 79.

[9] Young VR, Scimshaw NS. The physiology of starvation. Sci Am 1971;225:14–22.

[10] Edington PW, Edgerton VR. Biology of physical activity. Boston, MA: Houghton Miffin; 1976. p. 1–352.

[11] Klabunde RE. Cardiovascular physiology concepts. 2011. http://www.cvphysiology.com/CAD/CAD009.htm.

[12] Doenst T, Nguyen TD, Abel ED. Cardiac metabolism in heart failure implications beyond ATP production. Circulation Res 2013;113:709–24.

[13] Mader SS. Biology. 9th ed. Mc Graw Hill; 2007.

[14] Mack S. How bacteria produce ATP energy without mitochondrion?. 2007. http://www.madsci.org/posts/archives/2007-11/1193962676.Mi.r.html.

Chapter 11

Adenosine Triphosphate Synthase

Everything you do needs energy to power it, whether it is walking up a flight of stairs, eating dinner, fighting off an infection, or even just making new proteins for hair growth. The energy form used by all cells, from bacteria to man is adenosine triphosphate (ATP). Every process within a species, including DNA and protein synthesis, muscle contraction, active transport of nutrients, nerve activity, maintaining osmosis, or carbon fixation, requires a source of ATP.

Why is ATP such a special source of energy? ATP is a nucleoside triphosphate (ribose sugar, adenine base, and three phosphate groups), where a high-energy bond connects the third phosphate group to the molecule. This bond is unstable, and when it is hydrolyzed, it releases a substantial amount of free energy (~7 kcal/mol). As well as providing energy, ATP has multiple other critical roles within cells: it is one of the four nucleotides required for the synthesis of RNA (protein synthesis); it regulates many biochemical pathways; in mammals, it is released from damaged cells to elicit a pain response; and in photosynthetic organisms it promotes carbon fixation. However, since it is unstable and cannot be stored for long it is used for almost every conceivable process, each cell in the body must always be producing ATP to supply its needs. In total, an organism's requirement for ATP is substantial; the average human body generates over 100 kg of ATP each day (this is more than the average weight of an adult human). ATP synthase is the prime producer of ATP in cells, catalyzing the combination of adenosine diphosphate (ADP) with inorganic phosphate to make ATP.

Considering its fundamental importance in sustaining life, organisms evolved ATP synthase (ATPase) during the earliest stages of evolution, making it one of the oldest of enzymes; predating the photosynthetic and the oxidative phosphorylation enzyme mechanics. As such, ATPase has remained a highly conserved enzyme throughout all kingdoms and species of life. The ATPases found in the thylakoid membranes of chloroplasts and in the inner membranes of mitochondria in eukaryotes retain basically the same structure and function as their enzymatic counterparts in the membranes of bacteria. In particular, the subunits that are needed for catalysis show striking homology between species.

ATPases are membrane-bound ion channels (or actually transporters) which couple ion movement through a membrane with the synthesis or hydrolysis of a

Biology of Life. http://dx.doi.org/10.1016/B978-0-12-809685-7.00011-3

nucleotide, usually ATP. Different forms of membrane-associated ATPases have evolved over time to meet specific demands of cells. These ATPases have been classified as F-ATPase, V-ATPase, A-ATPase, P-ATPase, and E-ATPase based on functional differences. They all catalyze the reaction of ATP synthesis and/ or hydrolysis.

The driving energy for the synthesis of ATP is the H^+ gradient, while during ATP hydrolysis the energy from breaking the ATP phosphodiester bond is the driving force for creating an ion gradient. Structurally these ATPases can differ, F-, V-, and A-ATPases are multisubunit complexes with a similar architecture and possibly catalytic mechanism, transporting ions using rotary motors. The P-ATPases are quite distinct in their subunit composition and in the ions they transport, and do not appear to use a rotary motor.

The F-ATPases (for '*phosphorylation Factor*', also known as H^+-transporting ATPases) are the prime enzymes used for ATP synthesis, and are very much conserved throughout evolution. They are found in the plasma membranes of bacteria, in the thylakoid membranes of chloroplasts, and in the inner membranes of mitochondria of eukaryotes. These membrane proteins can synthesize ATP using a H^+ gradient, and work in reverse to create a H^+ gradient using the energy gained from the hydrolysis of ATP.

The V-ATPases (for '*Vacuole*') are found in the eukaryotic endomembrane system [vacuoles, Golgi apparatus, endosomes, lysosomes, clathrin-coated vesicles (transport external substances inside the cell)], plant tonoplasts, the plasma membrane of prokaryotes, and certain specialized eukaryotic cells. V-ATPases hydrolyse ATP to drive a proton pump, but cannot work in reverse to synthesize ATP. V-ATPases are critical to a variety of vital intracellular and intercellular processes such as receptor mediated endocytosis, protein trafficking, active transport of metabolites, homeostasis, and neurotransmitter release.

The A-ATPases (for '*Archaea*') are found exclusively in Archaea and have a similar function to F-ATPases (reversible ATPases), even though structurally they are closer to V-ATPases. A-ATPases may have arisen as an adaptation to the different cellular needs and the more extreme environmental conditions faced by Archaeal species.

The P-ATPases are found in bacteria and in eukaryotic plasma membranes and organelles. P-ATPases function to transport a variety of different compounds, including ions and phospholipids, across a membrane using ATP hydrolysis for energy.

The E-ATPases (for '*Extracellular*') are membrane-bound cell surface enzymes that have broad substrate specificity, hydrolyzing other NTPs besides ATP, as well as their most likely substrates are ATP, ADP. The F-ATPases can either produce ATP by harnessing the energy from a proton gradient, or they can work in reverse to create a gradient from the hydrolysis of ATP. F-ATPases provide a type of transporter for H^+ ions to pass through the membrane, and possess a unique rotary motor that couples the flux of ions with the enzymatic synthesis or hydrolysis of ATP.

In order to synthesize ATP, F-ATPases must capture the energy from the flux of protons through the ATPase channel. A high concentration of H^+ ions on one side of the membrane causes the H^+ ions to travel through the ATPase channel to the other side of the membrane where the H^+ ion concentration is lower. The flux of H^+ ions across the membrane drives the synthesis of ATP from ADP by F-ATPases.

Mitochondria contain two membranes: an outer membrane and a highly folded inner membrane, with a small intermembrane space between them. The center of the mitochondria is called the matrix. H^+ ions move from the intermembrane space to the mitochondrial matrix, because it is easier to accumulate a high concentration of ions in the small space between the two mitochondrial membranes than it is to fill the large central matrix, the latter take in more H^+ ions to create a gradient. In addition, the inner membrane is highly folded to provide a greater surface area for ATP synthesis. The H^+ gradient in mitochondria is produced during the tricarboxylic acid cycle in the digestion of glucose, where H^+ ions are stripped from the breakdown products of glucose and carried by to the intermembrane space. These H+ ions can only pass through the membrane via the ATP synthase channel, which harnesses this energy to make ATP.

By a similar mechanism, the H^+ ions in chloroplasts travel from the lumen through the ATPase channel in the thylakoid membrane to the stroma, and are supplied by the carrier NADPH, which in turn picks them up from the splitting of water using light. In bacteria, the H^+ ions travel from the space between the plasma membrane and the cell wall, through the ATPase channel in the plasma membrane to the cytoplasm, but the main function in some bacteria is ATP hydrolysis rather than synthesis.

F-ATPases are comprised of a soluble portion known as the F_1 ATPase complex (enzyme activity), which consists of five subunits (a, b, d, e, and g), and a membrane-embedded portion known as the F_0 ATPase complex (proton channel), which consists of at least three subunits (A, B, and C). In mitochondria, the F_0 complex usually has nine subunits (A-G, F6 and F8). There are some minor differences between the smaller subunits of F-ATPases found in bacteria, chloroplasts, and mitochondria. The F_1 ATPase complex is responsible for performing ATP synthesis or hydrolysis, while the F_0 ATPase complex makes the proton channel for the translocation of H^+ ions across the membrane. The F_1 complex contains three a subunits, three b subunits, and one of each of the other subunits, where the three b-subunits are catalytic and the three a-subunits are regulatory in function. There is a substrate-binding site on each of the three a-subunits and the three b-subunits, but only those on the b-subunits are active sites, while those on the a-subunits are regulatory sites.

During enzyme catalysis by the F_1 complex, some of the subunits rotate relative to the rest of the enzyme, making ATPase the smallest rotary motor known. In total, ATPase contains two rotational motors, one in F_1 (e and g subunits) driven by ATP hydrolysis, the other in F_0 (C subunit) promoted by the H^+ gradient, which are joined together so that the rotation of the two motors is coupled back-to-back. The two motors try to rotate in opposite directions, but the F_0 motor is usually stronger, using the force from the H^+ gradient flux to push the F_1 rotary motor in reverse, in order to drive ATP synthesis.

The reaction catalyzed by F-ATPase is very much reversible, such that ATP hydrolysis can be used to make a H^+ gradient by the reversal of the ion flux. In this case, the F_1 rotary motor works in a forward motion to hydrolyze ATP, and to drive the F_0 motor in reverse to create a H^+ gradient. The generation of a H^+ gradient can then be used to create ionic balance, or for active transport to drive substrate accumulation.

FURTHER READING

[1] Kagawa Y, Racker E. Partial resolution of the enzymes catalyzing oxidative phosphorylation. 8. Properties of a factor conferring oligomycin sensitivity on mitochondrial adenosine triphosphatase. J Biol Chem 1966;241:2461–6.

[2] Mccarty RE. A plant biochemist's view of H^+-ATPases and ATP synthases. J Exp Biol 1992;172:431–41.

[3] Fernández-Morán H, Oda T, Blair PV, Green DE. A macromolecular repeating unit of mitochondrial structure and function. Correlated electron microscopic and biochemical studies of isolated mitochondria and submitochondrial particles of beef heart muscle. J Cell Biol 1964;22:63–100.

[4] Gresser MJ, Myers JA, Boyer PD. Catalytic site cooperativity of beef heart mitochondrial F1 adenosine triphosphatase. Correlations of initial velocity, bound intermediate, and oxygen exchange measurements with an alternating three-site model. J Biol Chem 1982;257(20):12030–8.

[5] Nakamoto RK, Scanlon JAB, Al-Shawi MK. The rotary mechanism of the ATP synthase. Arch Biochem Biophys 2008;476:43–80.

[6] Doering C, Ermentrout B, Oster G. Rotary DNA motors. Biophys J 1995;69:2256–67.

Section IV

4.5 Billion Years of Evolution

Picture taken by Laurence Cole. Rough and tough street lass in Rawalpindi, Pakistan, 2001.

Section IV

4.5 Billion Years of Evolution

Chapter 12

Evolution Timeline

Here we try and explain evolution in its full magnitude, from the start of chemical life through to humans. The evolution tree that results is massive covering eons of time. It is best covered with a massive table. The dating of the table is in accordance with the time periods laid down by scientists: the Hadean Eon, 4600–4000 million years ago (MYA); the Archean Eon, 4000–3000 MYA; the Proterozoic Eon, 3000–540 MYA; the Paleozoic Era, 540–250 MYA; the Mezozoic Era, 250–66 MYA; and the Cenzoic Era 66 MYA to present (Table 12.1).

The following sources of information were sought and pooled for this table:

Timeline of the evolutionary history of life. https://en.wikipedia.org/wiki/Timeline_of_the_evolutionary_history_of_life
Exploring life's origins. http://exploringorigins.org/timeline.html
Evolution and geological timelines http://www.talkorigins.org/origins/geo_timeline.html
Timeline: The evolution of life https://www.newscientist.com/article/dn17453-timeline-the-evolution-of-life/
Tree of life rewritten: Evolutionary timeline looks more like a giant lollipop - and reveals new species appear every 2 million years http://www.dailymail.co.uk/sciencetech/article-2982564/Tree-life-rewritten-Evolutionary-timeline-looks-like-giant-lollipop-reveals-new-species-appear-2-million-years

It is all very complicated. One massive tree stretches in multiple direction. The key events are:

Formation of planet Earth 4.54 billion years ago
for the last 3.6 billion years, simple cells (prokaryotes);
for the last 3.4 billion years, cyanobacteria performing photosynthesis;
for the last 2 billion years, complex cells (eukaryotes);
for the last 1.2 billion years, eukaryotes which sexually reproduce
for the last 1 billion years, multicellular life;
for the last 600 million years, simple animals;
for the last 550 million years, bilaterians, water life forms with a front and a back;
for the last 500 million years, fish and proto-amphibians;
for the last 475 million years, land plants;

Biology of Life. http://dx.doi.org/10.1016/B978-0-12-809685-7.00012-5

TABLE 12.1 Evolution, Earth's Creation–Present

Hadean Eon (4600–4000 MYA)

Date	Event
4567 MYA	The solar system is formed.
4540 MYA	Planet Earth forms from planetesimal deposits and solar debris.
4450 MYA	The Moon is made when planet Earth collides with smaller planets.
4450 MYA	Initial atmosphere is hydrogen and helium, this then escapes Earth's gravity. A new atmosphere from volcanic emissions is carbon monoxide, carbon dioxide, water vapor, methane, hydrogen, and ammonia.
4280 MYA	Water lands on Earth from meteors and comets, and oceans are formed.
4000 MYA	Bombardment of Earth by meteors and comets stops.

Archean Eon (4000–3000 MYA)

Date	Event
3900 MYA	The Earth's crust solidifies.
3800 MYA	Chemical life and RNA molecules appear.
3500 MYA	Cyanobacteria, blue-green algae, shown to have existed.
3200 MYA	Early acritarchs shown to have existed.
3000 MYA	Advanced photosynthesizing cyanobacteria shown to have existed; producing oxygen as a waste product. The oxygen in the atmosphere slowly rises.

Proterozoic Eon (3000–540 MYA)

Date	Event
2800 MYA	It was not until the Earth was 1.7–2.8 billion years old, that a new atmosphere, rich in oxygen, emerged.
2800 MYA	Land continents formed from raised plates known as the Precambrian shields.
2400 MYA	It was not until the Earth was 1.7–2.4 billion years old, that a new atmosphere, rich in oxygen, emerged. Oxygenation of the atmosphere was led by cyanobacteria's oxygenic photosynthesis, causing the displacement of the carbon monoxide, carbon dioxide, water vapor, methane, and ammonia atmosphere.
2200 MYA	Organisms with mitochondria capable of aerobic respiration appeared.
2000 MYA	Diversification of acritarchs.
1850 MYA	Eukaryotic cells appeared. Eukaryotes contain membrane-bound organelles with diverse functions, probably derived from prokaryotes engulfing each other via phagocytosis. An oxygen atmosphere was an incentive to eukaryotic life.
1850 MYA	Complex single-celled life appeared.
1800 MYA	Bacteriophage demonstrated.

TABLE 12.1 Evolution, Earth's Creation–Present—cont'd

Hadean Eon (4600–4000 MYA)

Date	Event
1600 MYA	Acritarchs demonstrated, interpreted as the first eukaryotes.
1400 MYA	Appearance of terrestrial cyanobacteria.
1200 MYA	Meiosis and sexual reproduction occur in single-celled eukaryotes.
1200 MYA	*Bangiomorpha pubescens*, arctic Canada, another early eukaryote. It is the earliest example of sexual reproduction and complex multicellularity.
1200 MYA	Simple multicellular organisms evolve, comprising colonies of limited complexity. Multicellular algae evolve.
1100 MYA	Earliest dinoflagellates.
1000 MYA	First vaucherian algae.
760 MYA	The supercontinent Rodinia splits apart creating more coastline with continental shelves.
750 MYA	First protozoa appear.
580 MYA	The phyla of animals starts to appear.
580 MYA	Soft-bodied organisms developed, jellyfish, Tribrachidium, and Dickinsonia appeared.
560 MYA	Earliest fungi developed.
550 MYA	Ctenophora (jellyfish), Porifera (sea sponges), and Anthozoa (corals and sea anemones) appear.

Paleozoic Era (540–250 MYA)

Date	Event
543 MYA	Evolution of animals with hard body parts.
535 MYA	Major diversification of life in the oceans: chordates, arthropods, trilobites, crustaceans, echinoderms, mollusks, brachiopods, foraminifers, and radiolarians.
525 MYA	Earliest graptolites appear.
510 MYA	Vertebrates appeared in the ocean.
510 MYA	First cephalopods and chitons appear.
505 MYA	Burgess Shale appear.
500 MYA	Appearance of first vertebrates.
485 MYA	First vertebrates appear with true bones.
485 MYA	The earliest evidence for terrestrial activity by animals, tracks are made by multiple ~50 cm sized, multiple leg animals, preserved in sandstone.

Continued

TABLE 12.1 Evolution, Earth's Creation–Present—cont'd

Hadean Eon (4600–4000 MYA)

Date	Event
458 MYA	Fossils of spores and banded tubes indicate the first presence of land plants. These plants evolved from green algae.
450 MYA	*Conodonts* and echinoids appear.
440 MYA	First agnathan fish appear: *Heterostraci, Galeaspida,* and *Pituriaspida.*
434 MYA	The first plants move onto land, having evolved from green algae along lakes.
430 MYA	Organisms such as liverworts, lichens, fungi, and moss-like plants appear.
430 MYA	Waxy coated algae begin to live on land.
420 MYA	Millipedes have evolved as the first land animals.
420 MYA	Earliest ray-finned fish appeared, trigonotarbid arachnids, and land scorpions.
410 MYA	First signs of teeth in fish. Earliest *nautilida, lycophytes,* and *trimerophytes.*
395 MYA	First lichens and stoneworts appear. Earliest harvestmen, mites, hexapods (springtails), and ammonoids appeared.
395 MYA	The first known tetrapods left tracks on land.
380 MYA	The first amphibians move onto land. Evolving from fish, their fins develop into legs.
375 MYA	Vertebrates with legs first appeared, such as Tiktaalik.
370 MYA	First trees appeared on land.
363 MYA	Large variety of insects roamed the land and took to the skies.
363 MYA	First sharks swam in the oceans.
363 MYA	Seed-bearing plants and forests flourished on land. Land flora was dominated by seed ferns.
360 MYA	First crabs appeared in oceans.
350 MYA	First large sharks, ratfishes, and hagfish appear in the oceans.
350 MYA	Fish begin to form legs from their fins and lungs from their gills. One fish, the Eusthenopteron evolved the limb bones in its fins that were later necessary for the transition to land.
340 MYA	Diversification of amphibians.
330 MYA	Further evolution of coniferous trees, which are distinguished by their seeds.

TABLE 12.1 Evolution, Earth's Creation–Present—cont'd

330 MYA	First *amniote* vertebrates appeared.
324 MYA	Synapsid vertebrates, the ancestors of mammals, appear on land.
320 MYA	Reptiles evolved from amphibians, distinguished by their eggs that could be laid and hatched outside of water.
305 MYA	Earliest *diapsid* appeared.
300 MYA	First reptiles appeared, the Synapsids and Dyapsids, the ancestors of crocodiles and dinosaurs.
300 MYA	Winged insects evolve. At this point, winged insects are the only life that flies.
280 MYA	Earliest beetles and weevils evolve.
280 MYA	Seed plants and conifers diversify while lepidodendrids and sphenopsids decrease.
251 MYA	The *Permian–Triassic extinction event* eliminates over 90–95% of marine species. Terrestrial organisms were not as seriously affected as the marine biota.
Mesozoic Era (251–200 MYA)	
245 MYA	Earliest ichthyosaurs appeared.
240 MYA	Increase in diversity of gomphodont, cynodonts, and rhynchosaurs.
240 MYA	Sea urchins (Arkarua) first appear.
235 MYA	Evolutionary split between dinosaurs and lizards.
230 MYA	Roaches and termites evolve.
225 MYA	Earliest dinosaurs appeared.
225 MYA	First teleost fish appeared in oceans.
225 MYA	First mammals appeared.
225 MYA	First bees appeared.
221 MYA	The first dinosaurs appear. They are small dinosaurs.
220 MYA	Seed-producing gymnosperm forests dominate the land.
220 MYA	First flies appeared.
220 MYA	First coelophysoid dinosaurs came about.
205 MYA	First turtles came about.
200 MYA	First evidence of mammals.
200 MYA	Earliest examples of ankylosaurian dinosaurs.
220 MYA	First crocodiles evolved.

Continued

TABLE 12.1 Evolution, Earth's Creation–Present—cont'd

195 MYA	First pterosaurs and first sauropod dinosaurs. Diversification in small ornithischian dinosaurs: heterodontosauridae, fabrosauridae, and scelidosauridae.
190 MYA	First lepidopteran insects (*Archaeolepis*), hermit crabs, modern starfish, irregular echinoids, corbulid bivalves, and tubulipore bryozoans. Development of sponge reefs.
170 MYA	Earliest salamanders, newts, cryptoclidids, and elasmosaurid came about.
165 MYA	First rays and glycymeridid bivalves appeared.
160 MYA	Multituberculate mammals (genus *Rugosodon*) appear in eastern China.
155 MYA	First blood-sucking insects (ceratopogonids), rudist bivalves, and cheilostome bryozoans came about.
155 MYA	Archaeopteryx, a possible ancestor to birds, appears.
150 MYA	First birds like Archaeopteryx appear.
130 MYA	Origin of flowering plants.
130 MYA	The rise of the angiosperms, flowering plants boast structures that attract insects and other animals to spread pollen.
120 MYA	Oldest fossils of heterokonts, including both marine diatoms and silicoflagellates.
120 MYA	The earliest known placental mammal, the mouse-sized eomaia appears.
116 MYA	First birds with beaks without teeth appear.
115 MYA	First monotreme mammals appear.
110 MYA	First hesperornithes, toothed diving birds. Earliest limopsid, verticordiid, and thyasirid bivalves appear.
106 MYA	*Spinosaurus*, the largest theropod dinosaur, appears.
80 MYA	First ants appear.
70 MYA	Multituberculate mammals increase in diversity. First yoldiid bivalves.
68 MYA	Tyrannosaurus, the largest terrestrial predator appears in the fossil record.
Cenozoic Era (66 MYA– present)	
66 MYA	The Cretaceous–Paleogene extinction event eradicates about half of all animal species, including mosasaurs, pterosaurs, plesiosaurs, ammonites, belemnites, rudist and inoceramid bivalves, most planktic foraminifers, and all of the dinosaurs excluding their descendants, the birds.
66 MYA	Earliest rodents appear.
63 MYA	Evolution of the creodonts, meat-eating carnivorous mammals.

TABLE 12.1 Evolution, Earth's Creation–Present—cont'd

60 MYA	Diversification of large, flightless birds.
60 MYA	Earliest true primates evolve.
60 MYA	Rats, mice, and squirrels appear.
55 MYA	Modern bird groups diversify (first song birds, parrots, loons, swifts, woodpeckers); first whale; earliest lagomorphs, armadillos; appearance of sirenian, proboscidean, perissodactyl, and artiodactyl mammals.
55 MYA	Rabbits and hares evolve.
52 MYA	First bats appear.
50 MYA	Peak diversity of dinoflagellates and nannofossils, increase in diversity of anomalodesmatan and heteroconch bivalves, brontotheres, tapirs, rhinoceroses, and camels.
40 MYA	Modern butterflies and moths appear.
40 MYA	Giant whales appeared in the fossil record.
37 MYA	First nimravid, false saber-toothed cats.
35 MYA	Grasses evolve from among the angiosperms.
30 MYA	Earliest pigs and cats appear.
25 MYA	First deer appeared.
20 MYA	Parrots and pigeons evolved.
20 MYA	Chimpanzee line evolves.
20 MYA	First giraffes, hyenas, bears, and giant anteaters.
6 MYA	Australopithecines diversify.
5 MYA	First hippopotami appear.
4 MYA	First modern elephants, giraffes, zebras, lions, rhinoceros, and gazelles appeared.
4 MYA	Development of hominid bipedalism.
2.7 MYA	Evolution of Paranthropus.
2.5 MYA	The earliest species of Smilodon evolve.
2 MYA	First members of the genus *Homo* appear in the fossil record.
2 MYA	Widespread use of stone tools among hominids.
1.7 MYA	Extinction of australopithecines.
1.6 MYA	*Homo erectus* appears.
0.8 MYA	Short-faced bears (*Arctodus simus*) appear.

Continued

TABLE 12.1 Evolution, Earth's Creation–Present—cont'd

0.6 MYA	Evolution of *Homo heidelbergensis*.
0.4 MYA	Hominids hunt with wooden spears and use stone cutting tools.
0.35 MYA	Evolution of *Homo neanderthalensis*.
0.2 MYA	*Homo sapiens* appears in Africa.
0.01 MYA	First permanent *H. sapiens* settlements.
0.01 MYA	*H. sapiens* learn to use fire to cast copper and harden pottery.
0.006 MYA	Writing is developed in Sumeria.

for the last 400 million years, insects and seeds;
for the last 360 million years, amphibians;
for the last 300 million years, reptiles;
for the last 200 million years, mammals;
for the last 150 million years, birds;
for the last 130 million years, flowers;
for the last 60 million years, the primates,
for the last 20 million years, the family Hominidae (great apes);
for the last 2.5 million years, the genus *Homo* (including humans and their predecessors);
for the last 200,000 years, anatomically modern humans.

Clearly, the human evolutionary line was simple cells, cyanobacteria, photosynthetic cyanobacteria, single cell eukaryotes, multicell complex eukaryotes, animals, water-based life (possibly fish), amphibians, reptiles, mammals, primates, hominids, and finally humans. What a long journey.

FURTHER READING

[1] Doolitte FW. Uprooting the tree of life. Sci Am 2000;282:90–5.
[2] Barton NH, Briggs DEG, Eisen JA, Goldstein DB, Patel NH. Evolution. Cold Spring Harbour Laboratory Press; 2007.
[3] Hall BK, Hallgrimsson B. Evolution. 4th ed. Jones and Bartlett Publishers; 2008.

Chapter 13

Evolution of Chemical, Prokaryotic, and Eukaryotic Life

Chemical, prokaryotic, early eukaryotic life was life without a brain, life without nervous senses, and life without feeling, thinking and nervous movement.

CHEMICAL LIFE

Section I of this book focuses on chemical life or how life may have started. Chemical life is nothing more than a bunch of critical chemical reactions occurring inside a primitive cell. By and large, life is not controlled by DNA controlled synthesis or by RNA controlled synthesis, it is just a number of uncontrolled fluke chemical reaction. Should this be looked at as life or living per se? The answer is, not really.

Chemical life seemingly advanced or evolved with multiple stages leading to early prokaryotic life. A proposition of possible advancement steps needed to advance to prokaryotic life is presented in Fig. 13.1. This is simply chemical life advancing to simple chemical life with RNA and RNA enzyme or ribozymes, advancing to a simple chemical life with random proteins or enzyme activities, advancing to a simple chemical life with an adenosine triphosphate energy pathway, advancing to a simple chemical life with photosynthesis, advancing to a simple chemical life with RNA/DNA protein coding, advancing to a simple chemical life with enzymatic reproduction, advancing to a prokaryotic life (Fig. 13.1).

No example of a simple chemical life exists today except new life-forms forming in a simple round bilayer vesicle. In this respect, it is a simple microscopic round object, like a pencil dot or grain of dust. It is strictly inferred that this was the earliest life-form.

PROKARYOTIC LIFE

Prokaryotes are a microscopic single-celled organism that has neither a distinct nucleus with a membrane nor other specialized organelles. Prokaryotes include the bacteria and archaea. Prokaryote life seemingly started just over 4 billion

Biology of Life. http://dx.doi.org/10.1016/B978-0-12-809685-7.00013-7

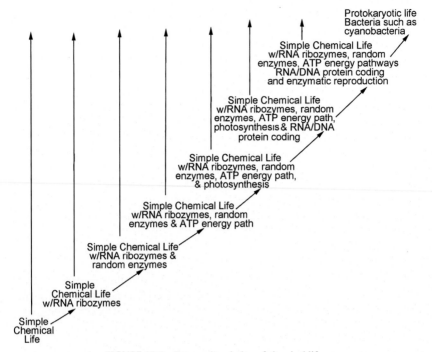

FIGURE 13.1   Proposed evolution of chemical life.

years ago, feeding off the early carbon dioxide, carbon monoxide, steam, nitrogen, hydrogen, and ammonia atmosphere.

Prokaryotes can be split into two domains, archaea and bacteria. In prokaryotes all the intracellular water-soluble components, proteins, DNA, and metabolites are located together in the cytoplasm enclosed by the cell membrane, rather than in separate cellular compartments. Bacteria do possess protein-based bacterial micro-compartments, which are thought to act as primitive organelles enclosed in protein shells. Some prokaryotes, such as cyanobacteria may form large colonies. Others, such as myxobacteria, have multicellular stages in their life cycles.

Prokaryotes have a cytoskeleton, which is much more primitive than that of eukaryotes. Besides homologues of actin and tubulin, the helically arranged building-block of the flagellum, flagellin, is one of the most significant cytoskeletal proteins of bacteria, providing structural backgrounds of chemotaxis, the basic cell physiological response of bacteria. At least some prokaryotes contain intracellular structures that can be seen as primitive organelles. Membranous organelles or intracellular membranes are common in some groups of prokaryotes, such as vacuoles or membrane systems devoted to special metabolic properties, such as photosynthesis. Some species also contain protein-enclosed microcompartments which have distinct physiological roles.

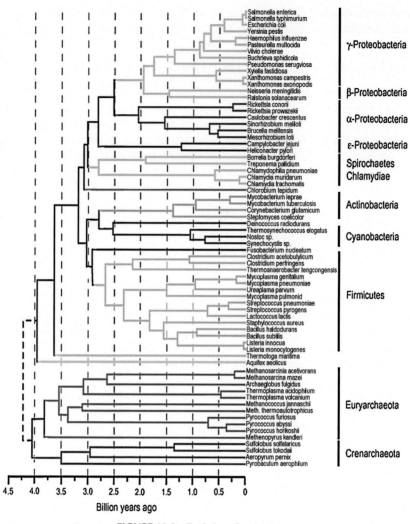

FIGURE 13.2   Evolution of prokaryotes.

Fig. 13.2 shows the evolution of prokaryotes starting 4.3–4.4 billion years ago and extending from a primitive carbon dioxide, carbon monoxide, steam, nitrogen, hydrogen, and ammonia atmosphere, through an oxygen-nitrogen atmosphere to the present. The evolved species listed in Fig. 13.2 included those causing common infections, *Salmonella enterica* (cause of food poisoning), *Treponema pallidum* (cause of syphilis), *Mycobacterium tuberculosis* (cause of tuberculosis), *Streptococcus pyrogens* (cause of sore throat) and *Streptococcus pneumonia* (cause of pneumonia).

Prokaryotic life is a very elementary life even though it has passed through billions of years of evolution or improvement. We kill bacteria by the billions with antibiotics and detergents with no care that we are killing life-forms. This is an elementary life-form that we kill and do not care about.

EUKARYOTE EVOLUTION FROM PROKARYOTES

The big difference between prokaryotes and eukaryotes are that prokaryotes do not have a clear nucleus and other specialized organelles. Eukaryotes came about by symbiosis. The evolution of eukaryotes probably took place after the development of the oxygen-nitrogen rich atmosphere, 2.8 billion years ago.

Amoebae have a symbiotic relationship with their guests, the bacteria that live inside their cell. Each amoeba and its independent bacterium are working together for mutual benefit, yet are separate organisms. Each bacterium or amoeba divides separately, gets its own energy, uses its own genes, and makes its own proteins. However, with their close relationship, it seems possible that after years of evolving together, these cells could evolve into a single integrated organism with a common set of genes and proteins. This is how eukaryotes came about, with the bacterium forming the mitochondrion in a cell.

Living species have evolved into three large clusters of closely related creatures, called domains: archaea, bacteria, and eukaryota. Archaea and bacteria are small, simple cells surrounded by a cell wall, with circular DNA containing their genes; these are prokaryotes. Virtually all the life we see today including plants and animals belongs to the domain eukaryota. Eukaryotic cells are more multifaceted than prokaryotes, the DNA is linear and found within a nucleus. Eukaryotic cells boast their own power plants, called mitochondria; these tiny organelles in the cell not only produce chemical energy, but also hold our understanding of the evolution of the eukaryotic cell. The eukaryotic cell ushered in a whole new era for life on Earth because these cells evolved into multicellular and multiorgan organisms.

How did the eukaryotic cell evolve? How did prokaryotic bacterium make this evolutionary leap to the more complex eukaryotic cell? The answer seems to be symbiosis or teamwork.

Evidence supports the concept that eukaryotic cells are the descendants of prokaryotic cells that joined together in a symbiotic unification. In fact, the mitochondrion itself is the descendent of a bacterium that was engulfed by another cell and ended up staying as a permanent part of that cell. The host cell profited from the chemical energy that the mitochondrion produced, and the mitochondrion benefited from the protected nutrient-rich environment surrounding it. This type of internal symbiosis, where one organism takes up permanent residence inside another and eventually evolves into a single lineage is called endosymbiosis (Fig. 13.3).

Biologist Lynn Margulis first made the case for endosymbiosis in the 1960s. Why should we think that a mitochondrion used to be a free-living organism in its own right? It turns out that much evidence supports this idea, most of all the many striking similarities between prokaryotes like bacteria and mitochondria.

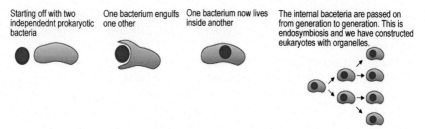

Starting off with two independednt prokaryotic bacteria

One bacterium engulfs one other

One bacterium now lives inside another

The internal baceteria are passed on from generation to generation. This is endosymbiosis and we have constructed eukaryotes with organelles.

FIGURE 13.3   Endosymbiosis.

Mitochondria have a cell membrane just like a prokaryotic cell does. Each mitochondrion has its own circular DNA genome just like a bacteria's genome. Viewed this way, mitochondria totally resemble bacteria living inside eukaryotic cells. The evidence for endosymbiosis applies not only to mitochondria, but to other cellular organelles also. Chloroplasts, for instance, are like tiny factories within plant cells that convert energy from sunlight into sugars. Evidence suggests that chloroplast organelles were also once free-living bacteria that entered cells through endosymbiosis.

The endosymbiotic events that generated mitochondria must have happened early in the history of eukaryotes, because all eukaryotes have them. Then, later, a similar event brought chloroplasts into eukaryotic cells, creating the lineage that led to plants.

Despite their many similarities, mitochondria and chloroplasts aren't living bacteria any more. The first eukaryotic cell evolved more than a billion years ago. Since then, these organelles have become completely dependent on their host cells. For example, many of the key proteins needed by the mitochondrion are imported from the rest of the cell. Sometime during their long-standing relationship, the genes that code for these proteins were transferred from the mitochondrion to its host's genome. Scientists consider this mixing of genomes to be the step at which the two independent organisms become a single individual.

EUKARYOTIC LIFE

Looking at history, the oxygen atmosphere occurred about 2.4 billion years ago. The first eukaryotic life-form appeared about 1.85 billion years ago. An oxygen atmosphere was an incentive to this first eukaryotic life-form. The first multicellular life-form and algae evolved about 1.2 billion years ago, and the first animal came about 0.580 billion years ago. Clearly, the first eukaryotic species were simple organisms and not multiorganed or brain-containing species like animals.

Multiple types of cells are found in eukaryotes. Animal cells are separate from other eukaryotes, most notably plant cells, lacking cell walls and chloroplasts. They also have smaller vacuoles. Due to the lack of a cell wall, animal cells can adopt a variety of shapes as needed in organs. There are many different types of cell in animals, there are, for instance, 210 distinct cell types in the adult human body.

Plant cells are quite different from other eukaryotic organisms. Their features include the following: a large central vacuole which maintains the cell's turgor and controls movement of molecules between the cytosol and sap; a primary cell wall containing cellulose, hemicellulose, and pectin, deposited by the protoplast on the outside of the cell membrane; the plasmodesmata linking pores in the cell wall that allow each plant cell to communicate with other adjacent cells; plastids, especially chloroplasts containing chlorophyll, the pigment that gives plants their green color; and flagellae (Bryophytes and seedless vascular plants lack flagella and centrioles except in the sperm cells. Sperm of cycads and ginkgo are large, complex cells that swim with hundreds to thousands of flagella. Conifers and flowering plants lack the flagella and centrioles that are present in animal cells).

Fungal cells are most similar to animal cells, with the following exceptions, fungal cells contain a cell wall that contains chitin. There is less definition between cells, ie, the hyphae of higher fungi have porous partitions called septa, which allow the passage of cytoplasm, organelles, and, sometimes, nuclei. Primitive fungi have few or no septa, so each organism is essentially a giant multinucleate supercell; these fungi are described as coenocytic.

Eukaryotes are a very varied group, and their cell structures are equally varied. Many have cell walls; many do not. Many have chloroplasts, derived from primary, secondary, or even tertiary endosymbiosis; and many do not. Some groups have unique structures, such as the cyanelles of the glaucophytes, the haptonema of the haptophytes, or the ejectisomes of the cryptomonads. Other structures, such as pseudopods, are found in various eukaryote groups in different forms, such as the lobose amoebozoans or the reticulose foraminiferans.

Fig. 13.4 shows the wide evolution of eukaryotes, and how animal are just a tiny part of the eukaryotic sphere or organisms. Eukaryotes can be classed as opisthokonts, the animal and fungus kingdoms together with the microorganisms that are conventionally assigned to the protist kingdom. Amoebozoa, a major taxonomic group containing about 2400 species of amoeboid protists, or amoeba and protozoa. Rhizaria, a species-rich supergroup of mostly unicellular eukaryotes. Plants, a group including all the flowers and plants found on Earth. Alveolates is a major groups of protists or single celled organisms. Heterokonts or stramenopiles are a major line of eukaryotes currently containing more than 25,000 known species. Most are algae, ranging from the giant multicellular kelp to the unicellular diatoms, which are a primary component of plankton. Discircristates clade consists of euglenozoa plus percolozoa. Excavates clade are a major subgroup of unicellular eukaryotes, containing a variety of free-living and symbiotic forms, and also includes some important parasites of humans.

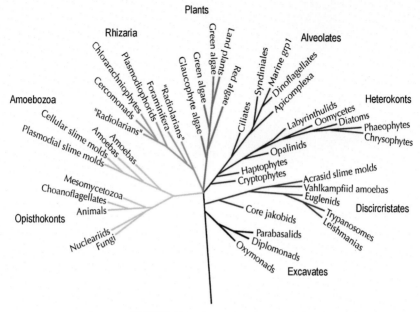

FIGURE 13.4   Eukaryote evolution.

FURTHER READING

[1] Whitman W, Coleman D, Wiebe W. Prokaryotes: the unseen majority. Proc Natl Acad Sci USA 1998;95:6578–83.

[2] Raven PH, Evert RF, Eichorm SE. Biology of plants. New York: W.H. Freeman; 1999.

[3] Dacks J, Roger AJ. The first sexual lineage and the relevance of facultative sex. J Mol Evol 1999;48:779–83.

[4] Scamardella JM. Not plants or animals: a brief history of the origin of Kingdoms Protozoa, Protista and Protoctista. Int Microbiol 1999;2:207–21.

Chapter 14

Animal Evolution, Small Brain and Advanced Brain

ANIMAL AND MAMMAL EVOLUTION

Selectively, chemical life evolved to prokaryotes and selectively through endosymbiosis prokaryotes led to eukaryotes. Through opisthokonts, and just one avenue of eukaryotic evolution channel of eukaryotes (Fig. 13.4) animals evolved with multiple organs. Then through animal evolution (Fig. 14.1), one avenue, the red avenue in Fig. 14.1, animals developed with brains and a central nervous system leading to birds, crocodiles, lizards, turtles, mammals, and other animals with a brain.

The brain was generally very small in animals, approximately 0.1% of total body weight, or one-thousandth of body weight. This brain was big enough to control the central nervous system, muscle response, and pain response with little space left to think and to reason. Yes, an animal could think without language "I am hungry, search for food," or "attack," or "let me find a place to pee," but not much more.

On multiple occasions evolution led to brain growth genes. But these could not be activated because the very inefficient placental system in most mammals, epitheliochorial placentation (Fig. 15.2) limited animal nervous tissue development.

One of the red avenues led to the evolution of mammals (Fig. 14.1), a group of mostly small brained animals. Mammals include the ancestors of primates and ancestors of humans (Fig. 14.2).

SMALL BRAIN ANIMALS

Mammals like sheep, horse, lemurs, and goats are small brain animals. As shown in the photo, they can smile when you say, "smile, say cheese" (Fig. 14.3). These animals live life to its fullest but have no understanding of life, what defines life, or that they will eventually die. Their tiny brain, approximately 0.1% or one-thousandth of their body weight, can coordinate the muscle contract signal or the pain signal from every nerve in the body, but has little brain space to think or plan.

Human beings possess a brain far beyond animals, 2.4% of body weight in size. Many ideas exist to explain the mind. Knowing the truth has vital implications!

Biology of Life. http://dx.doi.org/10.1016/B978-0-12-809685-7.00014-9

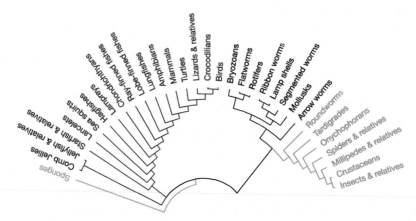

FIGURE 14.1   Animal evolution.

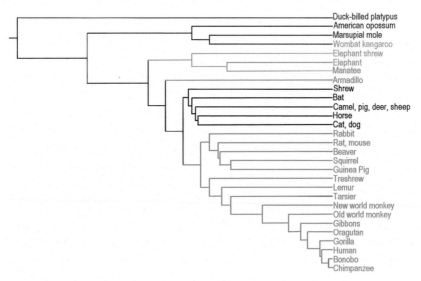

FIGURE 14.2   Evolution of mammals.

Biologists and religionists both agree that there is "something" about mankind that makes them different from animals. Yes, they can coordinate a nerve signal and a pain response, but they can think, plan, and coordinate speech.

Would it be good if an animal could think and plan? Theoretically if an animal could think out its plans and tactics for finding food in a language, it could probably do a much better job of finding food by knowing where to look. Evolution did not help these animals develop a bigger brain, solely because of the inefficiency of placentation in animals, epitheliochorial placentation, a mechanism too ineffective to support growth and development of expanding nerve tissues in animal fetuses.

FIGURE 14.3   Friendly face of mammals.

ADVANCED BRAIN ANIMALS

The dog, cat, and great apes, are examples of advanced brain animals, having a brain of more than 0.1% of body weight. The dog brain is 0.8% of body weight and the cat brain is 0.78% of body weight, these brains grew in size because of the development of zonary placentation, a slightly more efficient placentation than epitheliochorial placentation. Great ape brains also developed in size, the chimpanzee has a brain which is 0.8% of body weight and the orangutan has a brain which is 0.74% of body weight. This brain enlargement came about because of the development of the much more efficient hemochorial placentation in their ancestor, the lower simian primate. Chimpanzees, dogs and cats can think a little and remember a lot but cannot speak or plan because of the absence of the pertinent language areas of the brain needed for thinking.

It is thought that a chimpanzee may have a better short-term memory than humans do. In one experiment, when the number one through nine appeared randomly on a screen and then disappeared, the chimpanzees were able to recall the exact sequence and location of each number, drawing each number.

Dog memory can be best understood as primarily associative versus real memory. My dogs, Daisy, a female Rottweiler, and Cooper, a male German Shepherd remember places based on associations they have with people and places. If I put my Burberry cap on they know that we are going for a walk and get very, very excited, and start jumping up and down at the front door. They both walk beside me on a walk and know exactly where we are going, whether I take this route, that route, or the route around the housing estate. This memory will last for many years unless a new association to the Burberry cap is established by the dogs.

FURTHER READING

[1]   Cresswell Julia. The oxford dictionary of word origins. 2nd ed. New York: Oxford University Press; 2010. Having the breath of life', from animal air, breath, life.

[2]   Douglas AE, Raven JA. Genomes at the interface between bacteria and organelles. Phil Trans of Royal Soc 2003;358:5–17.

[3]   Alberts B, Johnson AL, Raff J, Roberts M, Walter K. Molecular biology of the cell. 4th ed. New York: Garland Science; 2002.

[4]   Ville CA, Walker WF, Barnes RD. General zoology. Saunders College Pub.; 1984. p. 467.

Chapter 15

The Evolution of Humans

The evolution of humans is an extremely complex subject, it is how humans came about through primates and hominids. An intricate subject is how the amazing, large, and complex human brain came about. It is the long process of how humans eventually evolved from mammals and the millions of mutations that must have linked them and how and why primates and hominids were critical intermediates. The evolution of humans is summarized in Fig. 15.1.

As shown in Fig. 15.1, the evolution of humans differs from the evolution of great apes or primates starting where we can say the path to humans started. It is my opinion that it started with the evolution of *Sahelanthropus tchadensis*, the root primate of humans and gorillas [1], and *Orrorin tugenensis* [2], the root primate of humans and chimpanzee. It must be realized that through multiple species steps *S. tchadensis* probably arose from the Gibbon, the common ancestor to gorillas, chimpanzees, and humans. These are shown as the start in Fig. 15.1. *Sahelanthropus tchadensis* was found in West-Central Africa. It was a partially bipedal primate. It had a small brain and small canine teeth. The oldest skeletons existed between 7 million years ago and 5.8 million years ago [1]. Findings are limited to skulls, however, *O. tugenensis* was a bipedal species that also walked on two legs in an upright manner with minimal canine teeth. The earliest *O. tugenensis* was found 5.8 million years ago. Both *S. tchadensis* and *O. tugenensis*, human ancestors, are today extinct species. How can we call them ancestors? We call them direct ancestors because they share features with us, such as, distinct bone markings and moldings, and genetic DNA sequences. This has been demonstrated and confirmed by hundreds of experts.

Stepping closer along the human branch of ancestors is *Ardipithecus ramidus*. The earliest *A. ramidus* was found 5.5 million years ago [3]. This species is also extinct. It shares clear traits with chimpanzees and gorillas but its features are shared most closely with humans. It was a forest-dwelling primate which walked on two feet and had minimal canine teeth [3]. Then came *Australopithecus afarensis* dated to 4 million years ago and now extinct. They had brains no larger than a chimpanzee brain but walked firmly upright on two legs and shared clear features with humans. They did not have canine teeth, like humans, and they were the first human ancestors to live on the savanna [4]. They were the

Biology of Life. http://dx.doi.org/10.1016/B978-0-12-809685-7.00015-0

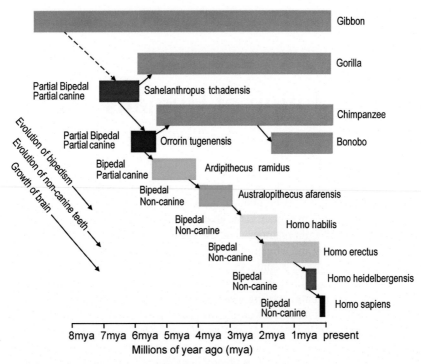

FIGURE 15.1 The evolutionary heritage of humans back to hominids and primates. *Orrorin tugenensis* is the ancestor of humans, chimpanzees, and bonobo. *Sahelanthropus tchadensis* is the ancestor of gorillas and humans.

first hominids to show the presence of a gene that causes increased length and ability of neurons in the brain, the SRGAP2 gene [5].

The earliest *Homo habilis* was dated to 2.5 million years ago but today is extinct [6]. The rather primitive attributes of this species have led experts to propose excluding *H. habilis* from the genus *Homo* and placing them instead in the genus of *Australopithecus* as *Australopithecus habilis*. *Homo habilis* was short and had disproportionately long arms compared to modern humans yet was a clear ancestor. *Homo habilis* had a cranial capacity less than half of the size of humans.

Homo erectus dates to approximately 1.8 million years ago [7]. *Homo erectus* was the first true hunter–gatherer ancestor, and also the first to have wandered out of Africa into Asia in large numbers [7]. It has a skeleton very structurally similar looking to humans but with a distinctly smaller brain capacity. Skeletons have been found as little as 70,000 years ago showing that they lived alongside humans but eventually became extinct.

Homo heidelbergensis dates back just 600,000 years ago in Africa and Europe [8]. *Homo heidelbergensis* is our grandfather or last species ancestor. It is an extinct species now, what happened to *H. heidelbergensis*? Why did it die off? It had a similar brain capacity to humans. *Homo heidelbergensis* lived in wooden huts, found today in Japan, and hunted with spears. *Homo heidelbergensis* was proven to have buried its dead. It was shown anatomically to have a partial sense of hearing and probably distinguished many sounds. It seemed to have some simplistic communication system. It was like a prehuman or our predecessor [8].

160,000 years ago, humans, *Homo sapiens*, evolved. Jewelry dating 100,000 years ago indicates that by then, people had developed complex speech and symbolism. Evidence suggests that 50,000 years ago, people began burying their dead ritually, they created clothes from animal hides, and they built simplistic traps to hunt animals. Approximately 10,000 years ago, it can be shown that agriculture was developed by humans, and that first villages and towns were built. This is how humans came about.

It is strange that all of our ancestors including *Homo neanderthalensis*, which developed side by side with humans, looked very similar to us, and had a slightly bigger brain than us, became extinct. *Homo neanderthalensis* became extinct 40,000 years ago, *H. erectus* became extinct 70,000 years ago, *H. heidelbergensis* became extinct 130,000 years ago, and so on. It appears that somehow we killed off our ancestor and our competitors as we kept going. Did this really happen? Or did we breed out our ancestors and competitors?

Scientists today suggest that human evolution is still ongoing, and has accelerated since the development of agriculture and civilization some 10,000 years ago. It is claimed substantial genetic differences exists between different current human populations [9]. The retention of lactase persistence into adulthood is an example of recent evolution.

THE HUMAN BRAIN

The large, highly complex human brain leading to human cognition, language, and intelligence is clearly the most distinguishing feature in humans, *H. heidelbergensis*, and *H. neanderthalensis* with its even bigger brain, and all other species. How did this come about? How did this evolve? How the brain evolved is a biochemical story that was first uncovered by Laurence A. Cole Ph.D., author of this book [10–12].

As thousands of species evolved, including hundreds of species of reptiles, fish, mammals, and aqueous mammals, evolution derived from umpteen types of spontaneous and other forms of mutations. Different size mammals evolved, two legged birds evolved, mammals with larger claws, larger mouths, and larger teeth all evolved. But, seemingly never did a large brain species, or a species with a more extensive nervous system evolve. This led paleontologists and archaeologists to question, how did humans ever evolve, or when did the large brain evolution occur, it clearly does not seem to happen?

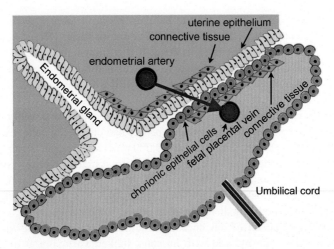

FIGURE 15.2 Epitheliochorial placentation, the *red arrow* shows the pathway of fetal nutrition.

Most mammals used a form of placental communication called epitheliochorial placentation. This is a generally very inefficient form of placentation (Fig. 15.2). Basically, the placenta loosely attaches to the uterus but does not physically implant or attach itself to the uterus. The nutrients have to get from the maternal uterine artery to a fetal placental vein. To achieve this, they have to leave the uterine artery and then somehow pass through connective tissue, then pass through uterine epithelial tissue, pass over the uterine space to the placenta, then pass through chorionic epithelial cells, then pass through placenta connective tissue and enter the placental cavity before reaching the fetal placental vein. This is very inefficient and few of the nutrients ever reach the fetus (Fig. 15.2). Mammals do produce brain growth expansion genes and growth expansion enzymes; these activities do not, however, work. The problem is that without an efficient means of obtaining nutritional supply in mammals, the sophisticated neuron tissue just does not grow. It is this problem with mammals that prevents any neuron or brain tissue growth or expansion.

For this reason, most epitheliochorial mammals have very limited brain growth, on average they have brains no bigger than 0.07–0.18% of body weight (Table 15.1). Yes, there are a few exceptions. Dogs and cats manage slightly improved brain growth (brain 0.78% and 0.8% of body weight, respectively) by adopting zonary placentation a slightly more efficient placentation technique (Table 15.1).

This is why there are few brainy mammals, nutritional efficiency in the fetus just will not support brain growth, even if genes are instructing brain growth. How did the brain growth story change to cause primate, hominid, and human brain growth (Table 15.1)?

Starting with primates, seven brain growth genes accumulated in species genomes. The brain growth genes and growth factors are MCPH1, ASPM, CD5RAP2, CENPJ, WDR62, CEP152, and STIL; genes accumulated among

TABLE 15.1 Brain Size of Epitheliochorial Placentation Mammals, Zonary Placentation Mammals, Early primates, Advanced Primates, Hominids and Humans With Hemochorial Placentation

	Brain % Body Weight
Epitheliochorial Mammals	
Lion	0.18
Horse	0.17
Elephant	0.18
Kangaroo	0.16
Camel	0.13
Giraffe	0.13
Cow	0.093
Pig	0.091
Zonary Mammals	
Dogs	0.8
Cats	0.78
Early Primates	
Prosimian primates	0.07
Advanced Primates	
Lower simian primate	0.17
Great ape	0.74
Hominids	
Homo habilis	1.2
Humans	2.4

primate and hominid species [13–20]. The only problem with the genes is that they could not function in the presence of inefficient epitheliochorial placentation. The early primates, for example, prosimian primates or lemurs utilized inefficient epitheliochorial placentation. These primates had tiny brains as a result of their inefficient placentation, 0.07% of body weight. It was with the evolution of the next species, lower simian primates, from prosimian primates that everything changed. The molecule chorionic gonadotropin (CG) was first produced in lower simian primates. It was this advance that initiated the evolution of the really larger brain species.

CG came about from a deletion mutation in the luteinizing hormone (LH) β-subunit gene. It placed an acidic C-terminal peptide extension on LH β-subunit on this newly evolved molecule CG. This increased the acidity and circulating half-life of the pituitary gonadotropin LH. This made the new hormone, CG, 8-fold more potent by increasing the acidity, which increased the circulating half-life from 0.3 to 2.4h [12]. LH was a molecule promoted by gonadotropin releasing-hormone (GnRH) in pituitary gonadotropic cells. The new more potent molecule was made in the lower simian primates by the pituitary, but was also made by placental cells that were likewise promoted by GnRH.

Interestingly, LH evolved from gonadotropin ancestral hormone-2 (GAH-2) found in fish, which evolved from an α-subunit ancestral hormone in fish, which evolved directly from transforming growth factor-β (TGFβ). As such, major elements of TGFβ, the 15 amino acid cysteine knot structure were present in LH and on the new molecule CG. The placenta had two types of cells, root cytotrophoblast cells and fused syncytiotrophoblast cells (made from fusion of up to 50 cytotrophoblast cells). The root cytotrophoblast cells produce an autocrine TGFβ receptor antagonist, the molecule that we call hyperglycosylated CG [21]. The fused cytotrophoblast cells, syncytiotrophoblast cells, produced a hormone similar to LH, the hormone that we call CG [22]. The hormone and autocrine together promoted synthesis and growth of all hemochorial placentation structural elements (Fig. 15.3) [23], the autocrine hyperglycosylated CG promoted the implantation of the hemochorial placentation placenta in the uterus [23].

Hyperglycosylated CG acted as an antagonist on the TGFβ receptor, blocking apoptosis and promoting cytotrophoblast cell growth, invasive collagenase and metalloproteinase production. This permitted cells to implant or bury themselves inside the uterus. Literally, the protease enzymes bored a hole into the uterus in which the blastocyst buried itself [23]. Implantation was mandatory for hemochorial placentation to work effectively. Hyperglycosylated CG promoted placental tissue growth and formation of the villous tissue that formed the core tree-shaped structure in the center of the hemochorial placentation chambers (Fig. 15.3). The CG promoted the fusion of the surface cytotrophoblast core tissue to make syncytiotrophoblast tissue [24]. The CG drove the extension of the maternal uterine spiral arteries to the hemochorial placentation tanks (Fig. 15.3) [25,26]. The CG also drove synthesis and development of the umbilical circulation to link the fetus to the villous tissues [27,28].

Lower simian primates now started implantation and much more efficient placentation or fetal nutrition, hemochorial placentation. Hemochorial placentation provided a much more efficient one cell barrier, the syncytiotrophoblast cell, between the maternal circulation and fetal circulation. Hemochorial placentation in lower simian primates, great apes, and humans is illustrated in Fig. 15.3. With the more efficient hemochorial placentation, the brain growth promoting genes started promoting brain growth beyond the minimum. With the evolution of CG, the brain size of lower simian primates enhanced from Prosimian, brain size 0.07% of body weight to lower simian primates, brain size 0.17% of body weight. This is how brain growth en route to humans started.

This was just the evolution of a first form of CG and hyperglycosylated CG that started brain growth on the road to humans. The lower simian primate molecule had just a small circulating half-life of 2.4h [12]. It was with many more mutations and increasing CG acidity in lower simian primates, great apes, *H. habilis*, and then in humans, that CG and hyperglycosylated CG were bumped up from weakly acidic molecules with five acidic sugar side chains, to molecules with six acidic sugar side chains, to molecules with seven acidic sugar side chains, to molecules with eight acidic sugar chains. With the addition of acidic sugars molecules of pI 6.3, circulating half-life 2.4h, were bumped up to

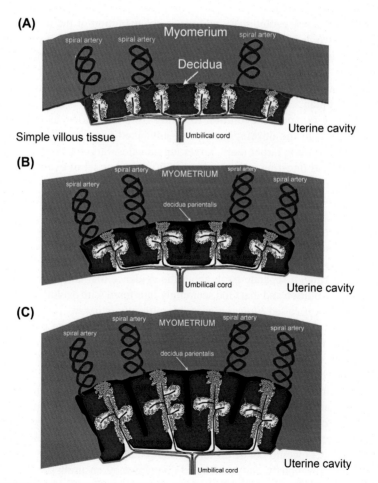

FIGURE 15.3 Hemochorial placentation in (A) lower simian primates, (B) great apes, and (C) humans.

more acidic molecules, pI 4.9, circulating half-life 6 h, to more acidic molecules (*H. habilis* is extinct, no blood data available), to the most acidic proteins made by humans, pI 3.5, circulating half-life 37 h. CG and hyperglycosylated CG are the most acidic proteins made in humans.

With this stepwise improvement and with increasing activity, hyperglycosylated CG promoted deeper implantation in the uterus of lower simian primates, great apes, *H. habilis*, and humans (Fig. 15.3) [12]. CG promoted increasing placentation potency in lower simian primates, great apes, *H. habilis*, and humans (Fig. 15.3). With increasing nutrition, brain promotion enzymes became more active and brain sizes grew from 0.17% of body weight (lower simian primates), to 0.74% of body weight (great apes), to 1.2% of body weight (*H. habilis*, available from skull), to 2.4% of body weight (humans) (Table 15.1). This is how the human brain came about.

There are a few evolutionary oddities in the evolution of the human brain. Firstly, the whole brain growth evolutionary plan from prosimian primates to lower simian primates, to great apes, to hominids, and then to humans seems so planned. It appears like some power or outside influence at the level of mammals or early primates seemed to plan this 40 million year brain evolution enlargement plan.

Furthermore, species evolution resulting from mutations, as dictated by Darwin, leads to the development of an advanced new species with larger claws, larger mouth, larger teeth, larger leg muscles, larger arm muscles, different diet, and intestine enzyme. About 200 different set of events separate advancing species. It appears that along the pathway of developing humans, repetitive mutation events occurred, causing brain expansion after brain expansion after brain expansion after brain expansion. Yet brain expansion never occurred previously. It really appears that an outside factor like God interfered and was involved. In Genesis, God claims to have created man according to his own plans.

I have written a separate book about the evidence that God was somehow involved in evolution and that God seemingly interfered with evolution to create man. The book is called *Eloivolution* and is currently *in press*.

I do not think that the story ends here. There is no evidence that humans are at the end of the evolutionary road. A more acidic variant of CG may easily arise next through mutations and the addition of further sugar side chains, and lead to an even bigger brained species. If it does not happen naturally through mutations and evolutions, it sure will happen unnaturally. I can buy milligrams of pure CG and synthesize large quantities of pure hyperglycosylated CG from the urine of choriocarcinoma patients. The processes of synthetically adding a further acidic sugar and linking it using glycosyltransferase to CG and hyperglycosylated CG would not be difficult. I could then administer these synthetic molecules intravenously, daily, to a new IVF pregnancy, ensuring super implantation, super development of the placenta, and the creation of a human with a much larger brain.

If you decide to try this, I want no association or responsibility; it will probably work, but may have complications. *Homo heidelbergensis* had the same size brain as humans. A species was needed to try the large brain size out and fix any neurologic connections. If we produce super humans, "*H. sapiens praesto*," we need to be sure that they would not be born with a major neurological defect. We know that there are seven brain expansion genes in humans, MCPH1, ASPM, CD5RAP2, CENPJ, WDR62, CEP152, and STIL, but we have not identified a brain angiogenesis gene, a gene that makes a hormone or enzyme that expands the circulation to accommodate the vastly extended regions.

The field of CG has complications. In humans, cancers steal that potent invasion pathway in human placental implantation, and use it to drive growth and invasion in malignancies [29]. Just as you can call CG and hyperglycosylated CG the molecules that drove brain growth in human evolution, you can also call CG and hyperglycosylated CG the molecules that promote invasion and growth in cancer [29]. Can you call them wicked molecules, I do not think so. They just drive cancers in people who insist on smoking, being exposed to carcinogenic substances, and who eat bad foods.

BIPEDALISM

The most critical part of human evolution, other than development of the brain, might be considered as bipedalism. Bipedalism is the switch from legs for transport to two legs for transport and two arms for tools, for lifting and moving, for writing, for driving, for computer operations, and thousands of other critical human functions.

Few modern species are bipeds whose principal method of locomotion is two-legged. Bipedalism has evolved multiple times, in the Triassic period some groups of archosaurs (ancestors of crocodiles) developed bipedalism, among their descendants are the dinosaurs, which were bipeds. Birds descended from dinosaurs. Many primate and bear species will adopt a bipedal gait in order to reach food or explore their environment. Several arboreal primates, such as gibbons, exclusively utilize bipedal locomotion during the brief periods they spend on the ground. Many animals rear up on their hind legs whilst fighting or copulating. A few animals commonly stand on their hind legs in order to reach food, to keep watch, to threaten a competitor or predator, or to pose in courtship, but do not normally move bipedally.

Limited and exclusive bipedalism can offer a species advantages: it raises the head, allowing a larger field of vision, with improved detection of dangers or resources; it allows access to deeper water for wading animals; it enables animals to reach higher food sources with their mouths; and while upright, non-locomotory limbs become free for other uses, including manipulation, flight in birds, digging, or combat.

There are many hypotheses as to how and why bipedalism evolved in humans. Bipedalism evolved well before the large human brain or the development of stone tools. Bipedal specializations are found in *Australopithecus* fossils from 5–3 million years ago. The evolution of bipedalism was accompanied by evolution in the spine including the movement in position of the foramen magnum. The different hypotheses are not necessarily exclusive and a number of forces may have acted together to promote human bipedalism. Possible reasons for the evolution of human bipedalism include freeing the hands for tool use, carrying, and food gathering.

According to the savanna-based theory, *A. afarensis* descended from the trees and adapted to life on the savanna by walking erect on two feet. The theory suggests that *A. afarensis* were forced to adapt to bipedal locomotion on the open savanna after they left the trees. This theory is closely related to the knuckle-walking hypothesis, which states that these human ancestors used quadrupedal locomotion on the savanna, as evidenced by characteristics found in *A. afarensis* forelimbs [30].

Other theories have been proposed that suggest wading in water and the exploitation of aquatic food sources or critical fallback foods may have exerted evolutionary pressures on human ancestors promoting adaptations which later assisted full-time bipedalism. It has also been thought that consistent water-based food sources had developed early hominid dependency and facilitated dispersal along seas and rivers.

Bipedal movement occurs in a number of ways, and require many mechanical and neurological adaptations. Energy-efficient means of standing bipedally involve constant adjustment of balance, and of course these must avoid overcorrection. The difficulties associated with simple standing in upright humans are highlighted by the greatly increased risk of falling in the elderly, even with minimal reductions in control system effectiveness.

Shoulder stability would decrease with the evolution of bipedalism. Shoulder mobility would increase because the need for a stable shoulder is only present in arboreal habitats. Shoulder mobility would support suspensory locomotion behaviors which are present in human bipedalism. The forelimbs are freed from weight bearing capabilities which makes the shoulder a place of evidence for the evolution of bipedalism.

Walking is characterized by an "inverted pendulum" movement in which the center of gravity vaults over a stiff leg with each step. Force plates can be used to quantify the whole-body kinetic and potential energy, with walking displaying an out-of-phase relationship indicating exchange between the two. Interestingly, this model applies to all walking organisms regardless of the number of legs, and thus bipedal locomotion does not differ in terms of whole-body kinetics.

In summary, the evolution of bipedalism and the evolution of the brain were possibly the two most critical part of human evolution. Overall, human evolution possibly started with the appearance of early simian primates such as the gibbon, that started approximately 40 million years ago in Africa. The development of the brain and bipedalism has been a continuous process since this time and may be still ongoing.

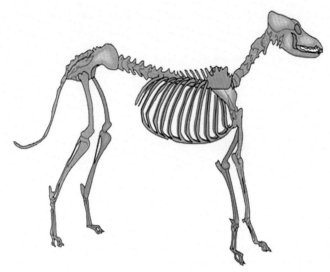

FIGURE 15.4 The skeleton of dogs.

It appears that the dog developed a somewhat larger brain with improved placentation methods. The dog brain in 0.8% of body weight or about one-third the size of the human brain (2.4% of body weight). I ask the question why was the pathway to an intelligent species started with primates? Maybe it was first started with wolves and dogs and then abandoned. Looking at the skeleton of dogs it is clearly a skeleton optimized in many ways, such as the design of the spine and pelvis, around the formation of a quadruped species (Fig. 15.4). Maybe this is why an interfering force may have abandoned it. It would have been difficult to evolve into a bipedal animal with arms that could be used for tools. The primates may have been an optimal group of species to adapt to bipedalism.

If anyone wants to know more you should read my most recently published research article on brain evolution [12], or contact me at larry@hcgab.com or 575-377-1330. You should also consider reading Hunt's comprehensive article on the evolution of bipedality [31]. You may want to read my lay book *Eloivolution, Discovery Institute Press, Seattle*, about hCG, evolution, and God, or read much more about CG and hyperglycosylated CG in the 414 page comprehensive book by myself and Stephen Butler [32].

REFERENCES

[1] Brunet B, Guy F, Pilbeam D, Mackaye HT, Likiu A, et al. A new hominid from the Upper Miocene of Chad, Central Africa. Nature 2002;418:145–51.

[2] Reynolds SC, Gallagher A. African genesis perspectives on hominin evolution. 2012. Cambridge Studies in Biolog Evolut Anthropol.

[3] White TD, Asfaw B, Yonas B, Yohannes H-S, Owen L, Gen S, et al. *Ardipithecus ramidus* and the paleobiology of early hominids. Science 2009;326:75–86.

[4] Raichlen DA, Gordon AD, Harcourt-Smith WEH, Foster AD, Haas W, Rosenberg K, editors. Laetoli footprints preserve earliest direct evidence of human-like bipedal biomechanics. PLoS One 2010;5:e9769.

[5] Reardon S. The humanity switch. New Scientist 2012;2012(2864):10–1.

[6] Richmond BG. Human evolution: taxonomy and paleobiology. J Anat 2000;197:19–60.

[7] Antón SC. Natural history of Homo erectus. Am J Phys Anthropol 2003;122:126–70.

[8] Mounier A, Marchal F, Condemi S. Is Homo heidelbergensis a distinct species? New insight on the Mauer mandible. J Human Evol 2009;56:219–46.

[9] Byars SG, Ewbank D, Govindaraju DR, Stearns SC. Natural selection in a contemporary human population. Proc Natl Acad Sci 2009;107:1787–92.

[10] Cole LA, Khanlian SA, Kohorn EI. Evolution of the human brain, chorionic gonadotropin and hemochorial implantation of the placenta: origins of pregnancy failures, preeclampsia and choriocarcinoma. J Reprod Med 2008;53:449–557.

[11] Cole LA. hCG and hyperglycosylated hCG in the establishment and evolution of hemochorial placentation. J Reprod Immunol 2009;82:111–7.

[12] Cole LA. The evolution of the primate, hominid and human brain. Primatology 2015;4:124.

[13] Keverne EB, Martel FL, Nevison CM. Primate brain evolution: genetic and functional considerations. Proc Royal Soc Lond 1996;262:689–96.

[14] Shi L, Ming L, Lin Q, Xuebin Q, Bing S. Functional divergence of the brain-size regulating gene MCPH1 during primate evolution and the origins of humans. BMC Biol 2013;62:1–11.

[15] Wang Y-Q, Bing S. Molecular evolution of microcephalin, a gene determining human brain size. Hum Mol Genet 2004;13:1131–7.

[16] Jackson AP, Eastwod H, Bell SM, Adu J, Toomes C, et al. Identification of microcephalin, a protein implicated in determining the size of the human brain. Am J Hum Genet 2002;71:136–42.

[17] Nicholas AK, Khurshid M, Désir J, Carvalho OP, Cox JJ, et al. WDR62 is associated with the spinal pole and is mutated in human microcephaly. Nat Genet 2010;42:1010–4.

[18] Bond J, Roberts E, Springell K, Lizarraga SB, Scott S, et al. A centrosomal mechanism involving CDK5RAP2 and CENPJ controls brain size. Nat Genet 2005;37:353–5.

[19] Guernsey DL, Jiang H, Hussin J, Arnold M, Bouyakdan K, Perry S, et al. Mutations in centrosomal protein CEP152 in primary microcephaly families linked to MCPH4. Am J Hum Genet 2005;87:40–51.

[20] Rouprina N, Pavlicek A, Mochida GH, Solomon G, Gersch W, et al. Accelerated evolution of the ASPM gene controlling brain size begins prior to human brain expansion. PLoS Biol 2004;2:653–64.

[21] Cole LA. Hyperglycosylated hCG, a review. Placenta 2010;8:653–64.

[22] Kovalevskaya G, Genbacev O, Fisher SJ, Cacere E, O'Connor JF. Trophoblast origin of hCG isoforms: cytotrophoblasts are the primary source of choriocarcinoma-like hCG. Mol Cell Endocrinol 2002;194:147–55.

[23] Cole LA. hCG and hyperglycosylated hCG, promoters of villous placenta and hemochorial placentation. In: Nicholson R, editor. Placenta: functions, development and disease. Nova Publishers; 2013. p. 155–66.

[24] Shi QJ, Lei ZM, Rao CV, Lin J. Novel role of human chorionic gonadotropin in differentiation of human cytotrophoblasts. Endocrinol 1993;132:387–95.

[25] Toth P, Li X, Rao CV, Lincoln SR, Sanfillipino JS, Spinnato JA, et al. Expression of functional human chorionic gonadotropin/human luteinizing hormone receptor gene in human uterine arteries. J Clin Endocrinol Metab 1994;79:307–15.

[26] Zygmunt M, Herr F, Keller-Schoenwetter S, Kunzi-Rapp K, Munstedt K, Rao CV, et al. Characterization of human chorionic gonadotropin as a novel angiogenic factor. J Clin Endocrinol Metab 2002;87:5290–6.

[27] Rao CV, Li X, Toth P, Lei ZM, Cok VD. Novel expression of functional human chorionic gonadotropin/luteinizing hormone receptor in human umbilical cords. J Clin Endocrinol Metab 1993;77:1706–14.

[28] Rao CV, Li X, Toth P, Lei ZM. Expression of epidermal growth factor transforming growth factor-alpha and their common receptor genes in human umbilical cords. J Clin Endocrinol Metab 1995;80:1012–20.

[29] Cole LA, Butler SA. B152 anti-hyperglycosylated human chorionic gonadotropin free β-subunit. A new, possible treatment for cancer. J Reprod Med 2015;60:13–20.

[30] Richmond BG, Strait DS. Evidence that humans evolved from a knuckle-walking ancestor. Nature 2000;404:382–5.

[31] Hunt KD. The evolution of human bipedality. J Human Evol 1994;26:183–202.

[32] Cole LA, Butler SA. Human chorionic gonadotropin (hCG). Burlington, MA: Elsevier; 2014. p. 414.

Chapter 16

Human Development

By far the biggest advance made by humans that has led to intelligence, and to thinking and planning is the development of communication, sounds and languages. Here we discuss the evolution of languages amongst humans.

LANGUAGE

We first examine the opinions of Christine Kenneally, published in her book "The first word: The search for the origin of languages" [1]. Multiple researchers believe that for a language to have evolved there had to be a necessity to speak, as opposed to having language evolve abruptly and then finding there was not much to talk about. In this opinion, language ability was not a coincidence, nor did it happen by a sudden enabling mutation for syntax. If linguistic ability evolved in the same way as other abilities, then the search for the origins of language includes a hunt for what was so advantageous about communication.

Animals that show advanced cognitive abilities suggest what the minds of our ancestors animals may have been like. Species that show complicated thinking have much in common with humans. Dolphins, elephants, and crows have long lives, extended infancy, complex systems of communication, and cultures where individuals have specialized roles. Female elephants, live years beyond their reproductive age and pass on knowledge to youths, such as how to interact with other elephants, as well as information about water holes or fruit trees.

Elephants used sounds like words. Researchers found that elephants produced distinct sounds for different reasons, like welcoming a member of the clan they had not seen for a while. Similarly, dolphins use decisive whistles, and seemingly named themselves with signature whistles. Whenever they met together they make a distinct sound. Dolphins also exchanged signature whistles when they separated.

Vervet monkeys make three dissimilar word-like alarm calls when they saw a threat from ground, tree, or air, cautioning their group to take fitting defenses. Some scientists believe that these alarm calls may be "protowords," a form of language. When animals make alarm calls, they are linking a sound to something in the world. All animals that researchers have studied can link a sound and a problem. Humans have built on this ability by using this ancient podium for sound and a specific problem, to evolve human language.

Biology of Life. http://dx.doi.org/10.1016/B978-0-12-809685-7.00016-2

Language is built in two levels. At the first level, sounds from the set encompassing the language, its phonemes, are combined into expressive sounds known as morphemes. Morphemes are found in birdsongs, in the songs of whales, and in other animal dialogues. The morphemes recur repeatedly, but no animal dialogues, not even birdsongs, has the range of sounds that humans have. At the next level, morphemes are combined into phrases that convey meaning. Linguists call this syntax.

Until recently, it was thought that only humans could make use of the rules of syntax, but different types of syntax have been found in the communication of many primate species, such as chimpanzees and gibbons. Gibbons arrange sounds into vocalizations that are different from those of other primates. These were used to communicate up to a kilometer in distance. What is important about these gibbon sounds is that the same set of sounds has two different meanings to a gibbon. The simple physical rules used by these primates in the wild deny the idea that creating meaning with defined rules, for example, subject, verb, and object is a human capability.

To better understand the capability that animals have for linguistic comprehension, efforts have been underway for decades to teach human communication to animals in cages, with amazing results. Chimpanzees and gorillas have gained a facility with American sign language to a degree that allows them to converse at the level of a young child. Kenneally states their achievements are on a par with children as old as four.

Sue Savage-Rumbaugh started educating language to bonobos in the 1970s. She claims that the best technique is to teach them indirectly, much as young children learn language, by having them hear it all around them. Young bonobos are raised in a language-rich environment by regularly being spoken to during feeding, playing, and grooming, rather than being drilled in repetitions of a word while pointing to the sign for it on a picture board. The result, Kenneally writes, is that the bonobos can, in addition to making simple declarations, converse about intensions and states of mind.

These bonobos demonstrated ingenuity in their use of words. For instance, some of them impulsively combined single words that they have learned to create new words, such as linking "water" and "bird" to form "waterbird," meaning a duck. Savage-Rumbaugh's star pupil, could join symbols to instruct what he wanted her to do. Savage-Rumbaugh has written that his immersion into a language-rich setting enabled him to acquire the mental mechanisms necessary for language acquisition.

The last common ancestor of humans and bonobos lived six million years ago. Savage-Rumbaugh demonstrates the ability to produce or understand language at the level of a young human child; it indicates that humans and their relatives inherited a set of skills that allowed them to make something like language as we know it. It makes sense to assume that our common ancestor evolved the trick of joining sounds to make meaning. Thus the basics of language existed long before the arrival of the genus, *Homo*.

A generation ago, it was generally accepted that language ability was located almost entirely in Broca's and Wernickes's areas on the left side of the brain (Fig. 16.1). This opinion was based upon research into the study of language problems in individuals whose brains were partly destroyed by trauma or disease. But according to Kenneally, this belief that the left brain is responsible for language is now superseded by the present understanding that language and other higher mental abilities are distributed throughout the brain.

To be sure, once the human brain has matured, language function is distributed across the brain and is not random. Particular areas take on parts of the task of perceiving, understanding, and producing language. But when the parts

Middle Paleolithic Period, 40,000-200,000 years ago
Homo sapiens and Homo neanderthalensis fboth flourished
They had stone axes
Hunting
Human burials
Ritual experiences

Upper Paleolithic Period, 10,000-40,000 years ago
Homo neaderthalensis became extinct
Cave art
Lived in houses
Seasonal group hunting and butchery
Food storage
Making clothing

Neolithic Period, 3,000-12,000 years ago
Farming and agricultural revolution
Farming of vegetables and animals
Community gatherings
Living as a communots in villages and towns
Trading
Centralization of administration
Division of labour

Civilization Period, Present-3,000 years ago
Centralization of governement and administration
Formation of cities
Community life
Community food supply
Social class structure
Organization of religions

FIGURE 16.1   Development of *Homo sapiens*.

of the brain that are best suited for language fail in young children, other parts can often assume the tasks of the destroyed regions, restoring a child's normal language level. This property of how brain tissue can be repurposed is called plasticity.

The plasticity of brain organization is not just a human property. The brains of apes change when the creatures have been moved from a laboratory setting to a more inspiring environment. An autopsy of a language-trained ape revealed it possessed a brain much larger than average for its species, and Savage-Rumbaugh maintains that this growth in size was the result of its learning experience. Her studies have concluded that primate brains are just as sensitive as humans to small increases in environmental complexity.

While apes in the wild do not sign to one another, and wild parrots do not talk each other, Kenneally suggests that brain plasticity accounts for their more sophisticated linguistic abilities in captivity, that is, the captive brains grow in response to the increased stimulation provided by constant human contact. There is a major difference, Kenneally writes, between being able to speak or not, but this difference result from circumstance and not from large differences in the brain.

She likewise dismisses differences in genetics that account for an innate grammar facility in humans. Much attention has been focused on the finding that a gene, called FOXP2, has a major affect on language ability. This has prompted speculation, that the evolution of this gene explains the development of language as a mutation. Since the emergence of language is parallel with the appearance of cognitively modern humans 50,000–80,000 years ago, presumably FOXP2 developed in its present form at that time.

Kenneally rejects this view for several reasons. The human version of FOXP2 is unique to *Homo sapiens*, but scientists have found that the gene is 98 percent the same in humans and songbirds. Mature birds with a greater expression of the gene tend to vary their songs more than others. This indicates that human communication depends on the same genetic foundations as communication by other animals.

As to the sudden appearance of FOXP2 as a gene for grammar, Kenneally states that Neanderthals, whose last common ancestor with *H. sapiens* lived at least 400,000 years ago, possessed a version of FOXP2 that was virtually identical to that found in humans. Whether Neanderthals could speak or not is unknown, but this is evidence that language ability evolved gradually in the same manner as other cognitive abilities. While some researchers claim that some genes may exist for the purpose of encoding grammar, Kenneally believes that this idea has been proven false.

The work on gesture has shown both a range that connects human and ape communication and significant changes between them. In our evolutionary history, some individuals must have been born with a greater ability to collaborate. These individuals were more fruitful and bred more offspring.

The gestural theory claims that human linguistics came about from gestures used for simple communication. Two types of evidence support this theory. Firstly, gestural language and vocal language depend on similar neural systems. Secondly, primates can use gestures for primitive communication, and some gestures resemble those of humans, such as the begging posture, with the hands stretched out, which humans share with chimpanzees [2].

Gestural language and vocal language depend on similar neural processes. The regions on the cortex that are responsible for mouth and hand movements border each other. Primates can use gestures or symbols for at least primitive communication, and some of their gestures resemble those of humans, such as the "begging posture," with the hands stretched out, which humans share with chimpanzees [2].

Research has found strong support for the idea that verbal language and sign language depend on similar neural processes. Patients who used sign language, and who suffered from a left-hemisphere problems, showed the same disorders with their sign language as vocal patients did with their oral language [3]. Other scientists found that the same left-hemisphere brain regions were active during sign language as during the use of vocal or written language [3].

HUMAN PSEUDOEVOLUTION

Humans evolved approximately 160,000 years ago. Humans directly descended from *Homo heidelbergensis*. The Middle Paleolithic period, 40,000–200,000 years ago, was the time when ancient humans, *H. sapiens* and *Homo neanderthalensis* appeared and flourished together. These peoples both used stone hand axes as tools to survive. The living method in the Middle Paleolithic period included scavenging, but there is also clear evidence of both hunting and gathering activities. Deliberate human burials, with evidence of ritual behavior are found during this period.

By 55,000 years ago, archaic humans were clearly tending to their elderly. Some evidence for cannibalism is found at this time. The Middle Paleolithic period finishes with the gradual disappearance of *H. neanderthalensis* for uncertain reasons and the ascendancy of *H. sapiens* about 40,000–55,000 years ago begins.

The Upper Paleolithic Age, 10,000–40,000 years ago, was a period of transitions (Fig. 16.1). The Neanderthals became extinct and modern humans started to have the world to themselves. Stone tools of the Upper Paleolithic Age were blade-based. Blades were then stone pieces twice as long as they are wide, with parallel sides. They were used to create an amazing range of tools with specific purposes. In addition, bone, antler, shell, and wood were used for both artistic and working tools, including the first eyed needles for making clothing (about 21,000 years ago).

The Upper Paleolithic Age is possibly best known for cave art, wall paintings, and engravings of animals and abstractions in caves (Fig. 16.1). People living during the Upper Paleolithic lived in houses with dugout floors, fireplaces, and windbreaks. Hunting became more focused, seasonal, and cultured, as shown by the remains of animals and selective butchery. Food storage was accomplished. Evidence suggests that small groups of people ventured on hunting trips and returned with meat to the base camps. The end of the Upper Paleolithic Age came about because of climate changes and global warming.

The Neolithic Revolution is often called the Agricultural Revolution (Fig. 16.1). It was the transition of human cultures from a lifestyle of hunting and gathering to a lifestyle of agriculture and settlement. Archaeological data shows that the farming of various types of plants and animals came about in many locations, starting around 12,000 years ago (Fig. 16.1).

The Neolithic Revolution involved much more than the adoption of a limited set of food-producing techniques. It would transform the small groups of hunter-gatherers that had dominated prehistory into groups based in villages and towns. These societies modified their environment by means of specialized food growth and employed food storage methods, permitting surplus food production. These developments generated the basis for densely populated settlements. It led to division of labor, trading economies, centralized administrations, property ownership, and fights over fertile properties.

Civilizations are societies organized into populated settlements. These are divided by social classes with a ruling elite and subordinate populations. The earliest appearance of civilizations is associated with the final stages of the Neolithic Revolution. The Neolithic revolution was reliant upon the development of grains, domestication of animals, and the growth of lifestyles which allowed the development of economies and the accumulation of surplus products.

Toward the end of the Neolithic period, many civilizations began to rise, starting around 3300 BCE with the growth of Egyptian society. The Iron Age around 1200 BCE saw many new societies emerge. Each society carefully investigated basic sciences and the principals of living and law, leading to what we call civilization. Religion and group religious practice is clearly part of civilization. The Jewish beliefs in God and communal religious practices led to civilization, as did the later Christian faiths and Moslem faiths. Yes, the Egyptian and Roman polygamous faiths aided the development of civilized societies, even though the polygamous religious beliefs were eventually abandoned (Fig. 16.1).

REFERENCES

[1] Kenneally C. The first word: the search for the origin of languages. New York: Penguin Books; 2008.
[2] Premack D, Premack AJ. The mind of an ape. New York: WW Norton and Co; 1983.
[3] Newman AJ, et al. A critical period for right hemisphere recruitment in American sign language processing. Nat Neurosci 2002;5:76–80.

Section V

Human Development

This section looks at the reproductive development of human life. If the reader wanted to examine human life and determine when human life really starts then there is a need to become familiar with: the physiology of the synthesis of spermatozoa; the physiology of the synthesis of the ova; the fertilization of a human embryo; embryo development; embryo implantation; and development of the human brain.

Picture taken by Laurence Cole. Child sitting on the porch of their jungle shed home, Papua New Guinea, 1999.

Chapter 17

Human Female Oogenesis

Three critical components of the menstrual cycle are: oogenesis or competitive generation of the dominant follicle and its oocyte; ovulation or expulsion of the oocyte from the ovary into the fallopian tube; and luteogenesis or formation of the corpus luteum from the follicular leftovers.

An ovarian follicle or emerging egg is composed of the germ cell or the oocyte and the surrounding granulosa and theca cells, the endocrine cells (Fig. 17.1). The follicle is the root structure for making the oocyte or germ cells. The granulosa and theca cells are the combined site of estrogen production (theca cells make androstenedione and granulosa cells aromatize androstenedione to make estradiol). Granulosa and theca cells hemorrhage and undergo differentiation to form corpus luteal cells (the site of progesterone production).

OOGENESIS

At first there are new primordial follicles developed by the action of follicle stimulating hormone (FSH). In the first stage of development, FSH promotes the growth of primordial follicles or primary oocytes (Fig. 17.2). FSH acts on a granulosa cell receptor.

About 15–20 primary oocytes race along the two ovaries to become the single dominant follicle each menstrual cycle. Occasionally, two dominant follicles can be generated leading to twin ovulation of oocytes, potentially leading to a nonidentical twin pregnancy. Rarely, triple dominant follicles are formed leading to triplet ovulations and to a nonidentical triplet pregnancy. New follicles are generated by the action of FSH. Unovulated primary follicles from a previous menstrual cycle can also be rescued from atresia and can join the primary follicle competition to become a dominant follicle (Fig. 17.1). Primary follicles are distinct having an oocyte and they do not have a central liquid space or antrum filled with liquor folliculi, this develops with secondary and advanced follicles.

In this first stage of follicular growth, FSH promotes the follicle to produce insulin-like growth factor (IGF)-1 and IGF-2. The IGF-1 and IGF-2 feedbacks to receptors on the follicle and promotes growth and development. With growth and development, the follicle becomes a secondary follicle with an antrum containing liquor folliculi secreted by granulosa cells (Fig. 17.2).

In the second stage of competition to become a dominant follicle, the rivalry becomes aggressive. FSH starts to promote insulin growth factor binding

Biology of Life. http://dx.doi.org/10.1016/B978-0-12-809685-7.00017-4

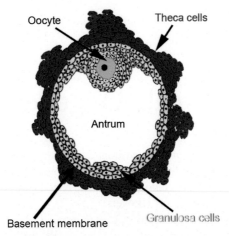

FIGURE 17.1 Structure of ovarian follicle.

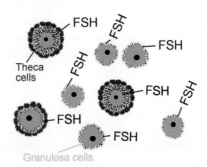

FIGURE 17.2 Primordial follicles. *Dark gray cells* are theca cells and *light gray cells* are granulosa cells. *White cells* are oocyte cells.

protein (IGF-BP). Every follicle regardless of its size produces the exact same amount of IGF-BP. IGF-BP binds the growth-promoting IGF-1 and IGF-2. The complex formed is immediately destroyed by a digestive protease, stopping the IGF-1 and IGF-2 from acting as growth promoters (Fig. 17.3). This empowers the competition. Because a large follicle with many FSH receptors will produce IGF-1 and IGF-2 over and above the amount bound by IGF-BP, so it has IGF-1 and IGF-2 left over to promote the follicles growth, and so fairs the best in the competition. (Fig. 17.3, Follicle 3). A small follicle may only produce the IGF-1 and IGF-2 which is 100% destroyed by the set amount of IGF-BP and the protease (Fig. 17.3, Follicle 1), so has no extra IGF-1 and IGF-2 left to promote the follicle further. Thus, the larger follicle is likely to win the dominant follicle race (Fig. 17.3, Follicles 2 and 3).

In the third stage of follicle growth and competition, inhibin produced by the granulosa cells promotes follicular growth (Fig. 17.4). The larger the follicle,

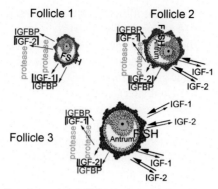

FIGURE 17.3 Formation of secondary follicle. *Dark gray cells* are theca cells and *light gray cells* are granulosa cells. *White cells* are oocyte cells. As illustrated, Follicle 1 is a small follicle not producing much insulin-like growth factor (IGF)-1 and IGF-2, all of the IGF-1 and IGF-2 is bound by the insulin growth factor binding protein (IGF-BP) and destroyed. Follicle 2 and Follicle 3 produce more IGF-1 and IGF-2 over and above the amount which is destroyed.

Third stage of oogenesis, inhibin competitive promotion

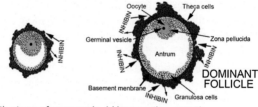

Fourth stage of oogenesis, LH promoting growth

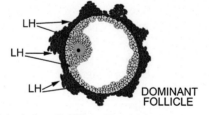

FIGURE 17.4 Action of inhibin and Luteinizing hormone (LH) on follicular growth.

the greater the amount of inhibin produced, the greater the inhibin promotion of follicle growth.

The three competitive stages, competitive FSH promotion, competitive destruction of IGF by IGF-BP, and competitive inhibin promotion, lead to the recognition of a dominant follicle between the two ovaries. This marks the finish of the follicular stage of the menstrual cycle, around day 11 of the menstrual cycle, and the start of the ovulatory phase of the menstrual cycle. In the ovulatory phase, a luteinizing hormone (LH) receptor now marks and appears on the granulosa cells and theca cells of the dominant follicle. In this fourth and last stage of oogenesis, the dominant follicle(s) are promoted to double in size by the rising LH or the day 14 LH peak (Fig. 17.4).

It is the dominant follicle which is promoted by inhibin to make the estradiol peak at around day 12 or 13 of the menstrual cycle. The estradiol peak promotes the LH and human chorionic gonadotrophin (hCG)–SO_4 peaks, which promotes the ovulation process (discussed later).

In mammals and in early prosimian primates, LH alone promotes steroid synthesis and LH alone promotes ovulation. Starting with lower simian primates a sulfated form of chorionic gonadotropin (CG–SO_4) evolved as a super active pituitary variant of LH. CG supplements all LH actions such as ovulation. In humans, hCG–SO_4 accounts for approximately half of the combined LH and hCG–SO_4 biological activity [1].

OVULATION

The ovulation process is started by the rising pituitary LH and hCG–SO_4 peaks. The estradiol peak promoted by inhibin at around day 11–12 of the average menstrual cycle increases the pulse frequency and the pulse amplitude of hypothalamus gonadotropin releasing hormone (GnRH). The GnRH pulses proceed through the hypothalamic–hypophyseal portal circulation to the gonadotrope cells, where high concentration of FSH, LH, and hCG–SO_4 are promoted. At this time, however, around 13–14 days of the average menstrual cycle, the granulosa cells of the dominant follicle(s) are producing the highest concentration of inhibin. This will inhibit pituitary FSH secretion by the pituitary gland. As such, just a peak of LH and hCG–SO_4 are made.

Apart from promoting a doubling in size of the dominant follicle, the rise of LH and hCG–SO_4 (both act on the same LH/hCG–SO_4 receptor), at around day 13–14 of the menstrual cycle, starts by firstly promoting meiosis in the oocyte. This is shuffling up the half set of genes from the father with the half set of genes from the mother and making haploid oocytes. Just the first stage of meiosis is completed in response to the LH oocyte signal, the second or final stage is completed during fertilization of the ovum (Fig. 17.5).

At the start, the oocyte is diploid like every other cell in the body, with one half of its 23 sets of chromosomes coming from the woman's mother and one half coming from the woman's father. LH and hCG–SO_4 then promote the first stage of meiosis. In meiosis, the diploid chromosomes are duplicated (prophase 1 leptotene) (Fig. 17.4), they are then randomly shuffled up (prophase 1 diakinesis), and then pulled apart to generate two cells (telophase). One of the telophase cells is discarded as the first polar body stored as waste in the dominant follicle, and the other becomes the uncompleted (until fertilization) haploid set in the oocyte and its expelled ovum.

After meiosis, the dominant follicle goes through LH and hCG–SO_4 promoted stigma formation and thinning, in preparation for ovulation. Stigma formation

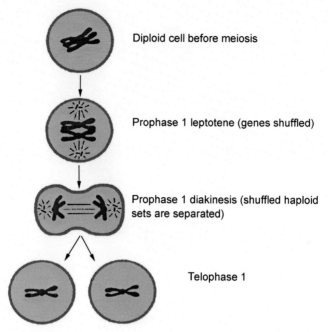

Diploid cell before meiosis

Prophase 1 leptotene (genes shuffled)

Prophase 1 diakinesis (shuffled haploid sets are separated)

Telophase 1

FIGURE 17.5   First stage of meiosis. There are 23 chromosome pairs but only 1 pair is illustrated. The *red genes* are from father and the *blue genes* are from the mother.

results in the formation of a bulge in the follicle at the likely site of ovulation. The walls of the bulge are made very thin to help ovulation (Fig. 17.6).

Then LH and hCG–SO$_4$ start ovulation. They promote production of three degradative proteases by the theca cells, a collagenase, a plasminogen activator, and a gelatinase (Fig. 17.7). These digest a hole in the theca and granulosa cells of the stigma of the follicle. As the follicular fluid squirts out of the hole in the follicle, the oocyte is expelled toward the fimbria of the fallopian tube (Fig. 17.7). The ovum then starts its journey down the fallopian tube.

LUTEOGENESIS

Following ovulation, the follicle is empty, but the theca and granulosa cell remain. A vacuum is made by expulsion of the follicular fluids. The empty follicle then collapses on the vacuum space left by the fluids (Fig. 17.8). Promoted by LH and hCG–SO$_4$, the membrane propia breaks down and blood vessels invade the empty follicle. The follicle hemorrhages and clots (Fig. 17.8). Promoted by LH and hCG-SO$_4$, luteogenesis is completed. The granulosa and theca cell hypertrophy, differentiate and form a corpus luteum. The corpus luteum becomes a yellow–orange color. The granulosa cells form the large superficial

FIGURE 17.6 Formation of stigma and thinning of the follicle.

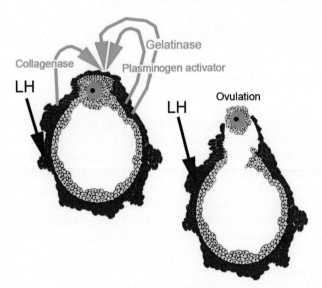

FIGURE 17.7 Ovulation.

lutein cells or outer crust of the corpus luteum (Fig. 17.8). The theca cells form the small lutein cells or inner mass of the corpus luteum.

The corpus luteum starts making progesterone under the control of LH and hCG–SO$_4$. The corpus luteum is present in the ovary from day 15 of the average 28 days menstrual cycle to day 26 of the menstrual cycle, when the ovary destroys it. During the luteal phase of the menstrual cycle, the corpus luteum starts managing inhibin production, blocking FSH production. The menstrual

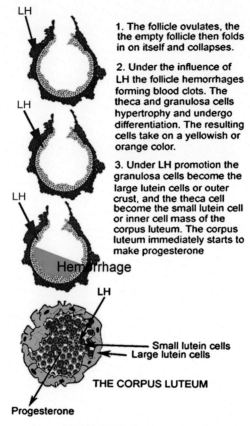

1. The follicle ovulates, the the empty follicle then folds in on itself and collapses.

2. Under the influence of LH the follicle hemorrhages forming blood clots. The theca and granulosa cells hypertrophy and undergo differentiation. The resulting cells take on a yellowish or orange color.

3. Under LH promotion the granulosa cells become the large lutein cells or outer crust, and the theca cell become the small lutein cell or inner cell mass of the corpus luteum. The corpus luteum immediately starts to make progesterone

FIGURE 17.8 Luteogenesis.

cycle finishes, and timing of the menstrual cycle starts, with a complete decline in progesterone production starting at around day 26 of a menstrual cycle, leading to shedding of the endometrial lining and hemorrhaging or menses.

REFERENCE

[1] Cole LA. Chapter 16. Pituitary sulfated hCG. In: Cole LA, Butler SA, editors. Human chorionic gonadotropin (hCG). 2nd ed. Burlington MA: Elsevier; 2014.

FURTHER READING

[1] Eckelbarger KJ. Oogenesis and oocytes. In: Purschke Günter, editor. Morphology, molecules, evolution and phylogeny in polychaeta and related taxa. 2005. p. 179–98.

Chapter 18

Human Male Spermatogenesis

This chapter is split into many sections. Firstly, spermatogenesis or the complex mechanism by which germ cells become sperm cells. Spermatogenesis is dealt with in three stages, mitotic proliferation, meiotic division, and cytodifferentiation and packaging of sperm cells. Altogether a massive amount of sperm is generated, approximately 300–600 spermatozoa are made per gram of testis per second, or multiple millions of sperms cell per day. Then, seminal fluids are considered, which are ejaculated along with the sperm.

MITOTIC PROLIFERATION

The germ cells of the immature testis are the source cells in the production of sperm. These germ cells are activated at puberty. These are the spermatogonial stem cells. With activation, these cells become A1 spermatogonia. Over 36–48 h these cells undergo multiple divisions or cloning steps, multiplying these cell many times over (A1–A5 spermatogonia). If cells undergo six divisions, for instance, each cell becomes 26 cells or is multiplied 64-fold (Fig. 18.1). After four to five divisions, the final clones become spermatogonia B and then resting primary spermatocytes (Fig. 18.1).

This mitotic division or cloning occurs in the basal compartment of the testicular seminiferous tubules (Fig. 18.4). During this fast cloning process nuclear division is always completed, but cytoplasmic division remains incomplete. As such, all spermatogonia remain joined together through their cytoplasm, joined by cytoplasmic bridges. This makes a spermatogonial syncytium or spermatogonia strings (see Fig. 18.2).

MEIOTIC DIVISION

The resting primary spermatocytes push their way into a compartment of the seminiferous tubules (Fig. 18.4). These cells undergo two stages of meiosis (first meiotic division and second meiotic division) (Fig. 18.3). All 23 chromosome pairs undergo meiosis. Each diploid cell with 23 pairs of chromosomes becomes, during meiosis, four haploid cells ready for use in sperm and for the joining with haploid chromosomes in the egg (Fig. 18.3). In the process of meiosis, each pair of chromosomes is first duplicated (generating four chromatids) in prophase 1-leptotene (Fig. 18.3), then tied together in prophase 1-zygotene

Biology of Life. http://dx.doi.org/10.1016/B978-0-12-809685-7.00018-6

135

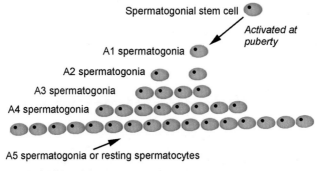

FIGURE 18.1 Cloning (mitotic proliferation) of spermatogonia.

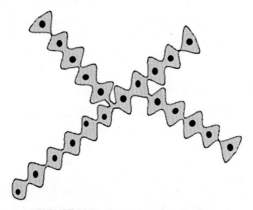

FIGURE 18.2 Spermatogonia syncytium.

(Fig. 18.3), and then undergoes interchange of genes between all four chromatids or "chiasmata," in prophase 1-diakinesis. The chromatids are then pulled apart in anaphase-1 to make two cells in late telophase 1 (Fig. 18.3). In Anaphase 2, all four chromatids are completely separated and make four cells each containing a single chromatid set telophase 2.

The product of meiosis is "early round spermatids." It can be calculated, that if a spermatogonia underwent six stages of duplication or cloning, it would generate 64 primary spermatocytes, after entering meiosis this would generate 256 early round spermatids. Considering losses due to failures, the actual number may be much less than this. All these spermatids continue linked like a syncytium as illustrated in Fig. 18.2. The round spermatids become separated.

Cytodifferentiation and Packaging of Sperm Cells

The most complex stage of sperm formation is the cytoplasmic differentiation and remodeling. This is done at the seminiferous tubule–Sertoli cell interface

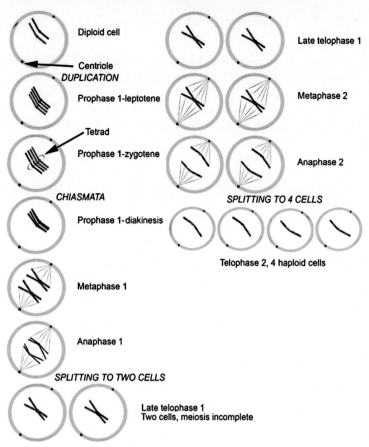

FIGURE 18.3 The two stages of meiosis. The example cells have one pair of chromosomes, in reality 23 diploid chromosomes undergo meiosis to make haploid cells.

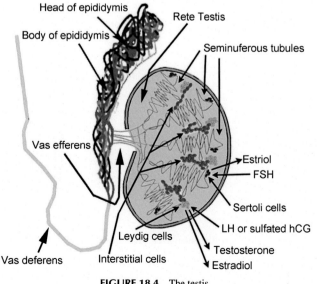

FIGURE 18.4 The testis.

(Fig. 18.4). Here, round spermatids are redesigned and a tail is added to give the spermatozoon propulsion. Firstly, out of glycoprotein rich granules which accumulate in the golgi apparatus, an acrosome granule forms growing over the nucleus forming a cap-like assembly (see Fig. 18.5 panel A and B). A single fiber flagellum or tail is formed that emanates from the acrosome and nucleus. The cytoplasm then becomes very squeezed and elongated. A neck or linking piece is formed by the cytoplasm, the nucleus and acrosome become flattened (Fig. 18.5 panel C). Nine fibers then form a long tail, forming the spermatozoon. Mitochondria accrue in the middle piece as a power center for tail propulsion.

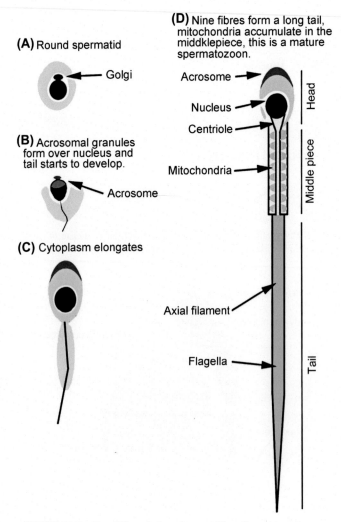

FIGURE 18.5 Cytodifferentiation of spermatids to form spermatozoon.

While mitosis or cloning occurs very quickly, these stages of cytodifferentiation and packaging of sperm cells are slower, taking approximately 45–50 days.

The flagella are full with dense fibers, which move with energy, adenosine triphosphate, generated in the mitochondria permitting spermatozoa propulsion. A mature spermatozoon is shown in Fig. 18.5 panel D. It distinctly has a head, a middle piece, and a trail or flagella. Rounds of spermatogenesis are initiated at constant time intervals in males.

Endocrine Control of Spermatogenesis

Spermatogenesis begins at puberty, when testosterone levels rise. Testosterone is critical to spermatogenesis. In the lack of testosterone, spermatogenesis only proceeds as far as the prophase 1-leptotene stage of meiosis (Fig. 18.3). Hypophysectomy or removal of the pituitary gland leads to an absence of luteinizing hormone (LH). With the absence of LH, Leydig cells stop producing testosterone and spermatogenesis comes to a halt. In this respect, LH is as critical to spermatogenesis as testosterone.

The role of follicle stimulating hormone (FSH) in men is less sure. FSH promotes growth of testosterone receptors on Sertoli cells and seminiferous tubules, this is important. Data with rodents advises that FSH binding Sertoli cells increases the number of spermatogonia or resting spermatocytes formed prior to meiosis. Basically, LH and testosterone are critical to spermatogenesis in men, but the role of FSH is secondary and less critical.

Maturation of Spermatozoa

Spermatozoa are released from Sertoli cells into a saline fluid and then washed toward the rete testis (Fig. 18.4). As the fluid passes through the rete testis the composition of the fluid changes, it becomes oxygen rich (needed to power middle piece mitochondria) and has serum albumin. The spermatozoa are then washed into the vas efferens and into the epididymis (Fig. 18.4). In the epididymis fluid is adsorbed and spermatozoa concentrated 100-fold. The epididymis then adds secretory goods including fructose, carnitine, and glycoproteins. The glycoproteins coats the surface of the spermatozoa. Passage through the epididymis takes 6–12 days and affects spermatozoa behavior profoundly. Spermatozoa in the epididymis have acquired what is needed to swim progressively by being supplied with nutrients, including fructose as sugar fuel, and by spermatozoa being coated with glycoprotein, or coated with charged sugars (sialic acid). This whole procedure of activation of spermatozoa is crucially dependent upon stimulation of the epididymis by testosterone.

In humans, the epididymis can stock sperm for 1–3 days. After the epididymis, spermatozoa enter the vas deferens (Fig. 18.4). The normal vas deferens serves as a storage compartment for spermatozoa. In the absence of ejaculation spermatozoa seep into the urethra and are washed away.

Seminal Fluids

Following the epididymis, spermatozoa enter the vas deferens (Fig. 18.4). The normal vas deferens serves as a storage compartment for spermatozoa. Ejaculated spermatozoa are carried to the female tract in seminal plasma. The combination of spermatozoa and seminal plasma are called semen. The seminal plasma offers nourishing factors such as fructose and sorbitol. It also provides an alkaline buffering capacity to neutralize the acidic vagina. Seminal plasma contains leukocytes to assist in sperm fertility.

The vas deferens first transports the spermatozoa to the seminal vesicles (Fig. 18.6). From the seminal vesicles the sperm passes into the prostate gland, and then through the Cowper's gland before entering the urethra and being ejaculated (Fig. 18.6). In the seminal vesicles, most of the seminal plasma is made. A protein call lipofuscin from dead epithelial cells give the semen a yellowish color. The seminal vesicles supply the semen with fructose and sorbitol as fuel, and an alkaline buffer to cope with the acidity of the vagina (Fig. 18.7).

In the prostate gland, semen gets citric acid as a calcium chelator. The semen is supplied with 1% proteolytic enzymes to burn the way of spermatozoa into ova. The semen also receives, in the prostate, an alkaline phosphatase to cleave glycerylphosphorylcholine for use in phospholipid metabolism.

The Cowper's gland provides the semen with a mucus lubricant that clears the flow of semen through the urethra. The viscous mucus produced in Cowper's gland is the principal component of the preejaculate in men.

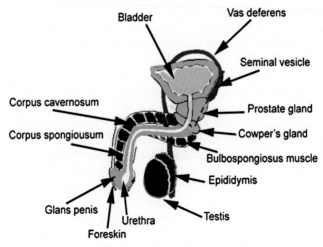

FIGURE 18.6 Male reproductive anatomy.

The Ingredient of Semen

1. Testis   The sperm, ~15 million cells per milliliter

2. Seminal vesicle   Lipofuscin, the yellow coloring of semen
                     Fuctose, an energy source or fuel
                     Sorbitol, an energy source or fuel

3. Prostate gland   Citric acid, a Ca^{++} chelator
                     Proteolytic enzymes, 1% w/w, to burn
                     entrance into ovum.
                     Alkaline phosphatase, to cleave gycerylph-
                     osphorylcholine for lipid metabolism.

4. Cowper's gland   Mucus lubricant, to clear flow through
                     urethra (principal ingredient of semen).

FIGURE 18.7   The components of semen.

FURTHER READING

[1] Spermatogenesis. https://en.wikipedia.org/wiki/Spermatogenesis.

[2] Wang C, McDonald V, Leung A, Superlano L, Berman N, Hull L, et al. Effect of increased scrotal temperature on sperm production in normal men. Fertil Steril 1997;68:334–9.

[3] Harrison RG, Weiner JS. Vascular patterns of the mammalian testis and their functional significance. J Exp Biol 1949;26:304–16.

[4] Carreau S, Bouraima-Lelong H, Delalande C. Role of estrogens in spermatogenesis. Front Biosci 2012;4:1–11.

[5] Díez-Sánchez C, Ruiz-Pesini E, Montoya J, Pérez-Martos A, Enríquez J, López-Pérez M. Mitochondria from ejaculated human spermatozoa do not synthesize proteins. FEBS Lett 2012;553:205–8.

Chapter 19

Sperm Activation, Fertilization, Morula, Blastocyst Formation, and Twinning

The semen has been through a meticulous synthesis and refinement process. In the epididymis, the sperm has been activated and in the prostate gland, proteolytic enzymes have been added to the semen to help penetration of the ovum; why is that not enough? The sperm now has to be capacitated in the uterus to activate it, to reach and search for the ovum, and then go through the acrosome reaction to activate the sperm before it can penetrate the ovum. Do we need to cook and then cuddle the sperm? When does sperm modification stop?

SPERM ACTIVATION

Vaginal intercourse leads to the release of semen into the vagina and uterus. That the sperm enters the vagina and uterus does not mean that it will propel its way to an ovum and fertilize it. If mature spermatozoa are incubated with oocytes in a test tube, fertilization either does not occur at all, or it takes many hours to complete. In contrast, if spermatozoa are removed from the vagina, uterus, or fallopian tubes 2 h after coitus, they are completely different and are capable, in a test tube, of immediate fertilization. These sperm have clearly been activated in some way in the uterus or fallopian tubes.

What we understand occurs to sperm on entering an estrogen-primed uterus is called sperm capacitation, which enhances sperm propulsion. Furthermore, the sperm cannot penetrate the zona pellucida or shell of an ovum without going through the acrosome reaction, a second form of activation needed for penetrating the ovum. Here we describe these two activation procedures.

SPERMATOZOA CAPACITATION

Semen enter the vagina and spermatozoa proceed through the uterus to the fallopian tube. The spermatozoa is propelled by the movement of the flagella or

Biology of Life. http://dx.doi.org/10.1016/B978-0-12-809685-7.00019-8

spermatozoa tail (Fig. 18.5). In order to be propelled properly the spermatozoa must be activated or capacitated. The capacitation process has two segments that superactivate the spermatozoa, boosting its propulsion and preparing it for entering into the oocyte vicinity (Fig. 19.1).

In the first segment of capacitation, a change occurs in the movement properties of the spermatozoa. The wave-like patterns of the spermatozoa's tail or flagella are made stronger, leading to wider amplitude flagella strokes or much more efficient movements in the uterus and fallopian tubes. Classically, a precapacitation spermatozoon appears motionless under a microscope, while a post-capacitation spermatozoon appears wiggling.

In the second segment of capacitation, a change occurs in the surface membrane, making spermatozoa responsive to signals encountered in the vicinity of the ovum. An acrosome reaction occurs, leading to increased calcium permeability and leading to a rise in intracellular calcium levels. In addition, Calmodulin-binding proteins are stripped from the spermatozoa surface, promoting calcium permeability. As a result of elevated calcium, adenyl cyclase activity within the spermatozoa rises. This results in cyclic adenosine monophosphate production and increased phosphorylation of proteins; Spermatozoa motility is enhanced accordingly (Fig. 19.1).

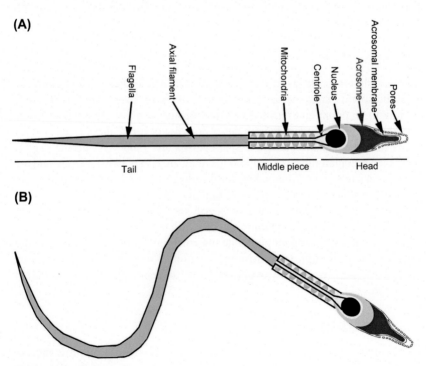

FIGURE 19.1   The spermatozoon and capacitation. (A) A noncapacitated spermatozoon and (B) A capacitated spermatozoon.

SPERMATOZOA ACROSOME REACTION

When sperm comes into contact with the shell of an ovum, the zona pellucida, the ovum's acrosome reaction occurs, to finally activate the spermatozoa. The acrosome reaction gives the spermatozoa the needed changes to burn its way across the zona pellucida and fertilize the ovum. The porous membrane stuck to the head of the spermatozoa swells in the acrosome reaction. The porous membrane is lost as it swells and fuses with the underlying acrosome membrane (Fig. 19.2). This reaction, like spermatozoa capacitation, leads to a large increase in intracellular calcium. An agent is secreted by the zona pellucida called ZP3, which activates the acrosome reaction. With the loss of the porous membrane, comes the release of multiple proteolytic enzymes such as hyaluronidase and acronsin, used for penetration of the zona pellucida by the spermatozoa in the fertilization process.

OVUM PROPULSION

The ovulated oocyte or ovum is surrounded by a zona pellucida cell shell and cumulus cells (Fig. 19.3). The ovulated oocyte is placed in the fimbriae of the fallopian tube and transported through the fallopian tube to the uterus. The mucosa of

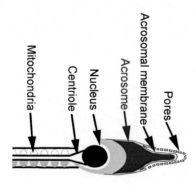

A head membrane of sticky pores is held on the spermatozoon. Upon contacting the zona pellucida of an ovum the Acrosome Reaction occurs.

The membrane of sticky pores is partly shed. During this time it is dropping behind the spermatozoon head. It is still attached to head throug its stickiness.

Finally all pores are shed exposing the forward part of the acrosome head. The Acrosome Reaction leads to release of proteolytic enzymes, hyaluronidase and acrosin, needed for penetration of the zona pellucida of the ovum.

FIGURE 19.2   Acrosome reaction.

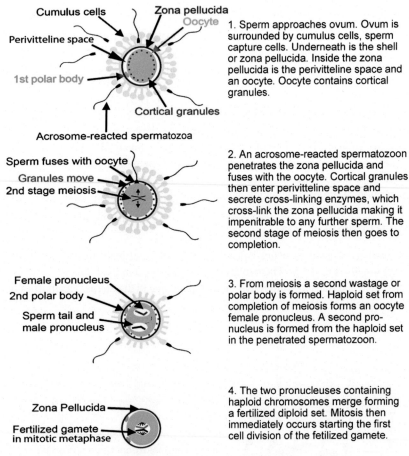

1. Sperm approaches ovum. Ovum is surrounded by cumulus cells, sperm capture cells. Underneath is the shell or zona pellucida. Inside the zona pellucida is the perivitteline space and an oocyte. Oocyte contains cortical granules.

2. An acrosome-reacted spermatozoon penetrates the zona pellucida and fuses with the oocyte. Cortical granules then enter perivitteline space and secrete cross-linking enzymes, which cross-link the zona pellucida making it impenitrable to any further sperm. The second stage of meiosis then goes to completion.

3. From meiosis a second wastage or polar body is formed. Haploid set from completion of meiosis forms an oocyte female pronucleus. A second pronucleus is formed from the haploid set in the penetrated spermatozoon.

4. The two pronucleuses containing haploid chromosomes merge forming a fertilized diploid set. Mitosis then immediately occurs starting the first cell division of the fetilized gamete.

FIGURE 19.3 The steps of fertilization.

the fallopian tube provides nourishment to the ovum or zygote (fertilized ovum). Secretions from the Peg cells in the fallopian tube provide the needed nourishment. The smooth muscle walls of the fallopian tube constricts and contracts providing a squeezing vacuum-like propulsion to the ovum, permitting the ovum to move through the fallopian tube length. This is promoted by prostaglandins and by estradiol. This effectively moves the ovum through the fallopian tube to be fertilized.

FERTILIZATION

After the acrosome reaction, the exposed acrosome becomes the head of the spermatozoon. This is covered in the acrosome reaction by proteolytic enzymes. These digest a path through the zona pellucida through which the spermatozoon passes. Penetration through the zona pellucida takes 5–20 min (Fig. 19.3). The head of the spermatozoon then fuses with the plasma membrane of the oocyte (ovum) (Fig. 19.3). Only spermatozoon that have undergone the acrosome reaction are capable of permeating the zona pellucida and fusing with the plasma membrane. Calcium from the spermatozoon is required for this fusing process. Once fusion has occurred, the spermatozoon freezes all movement. Within 1–3 min of fusion, a dramatic increase in free calcium in the oocyte occurs.

As soon as fusion occurs, the oocyte becomes a zygote. The zygote first prevents further spermatozoon from crossing the zona pellucida and fertilizing it. Upon fusing with the oocyte, cortical granules are released from the oocyte into the perivitelline space (Fig. 19.3). The cortical granules contain enzymes which cross-link the glycoproteins in the zona pellucida together. This make the zona pellucida impenetrable to further spermatozoon, sealing the oocyte. At the same time, the follicle now completes the second stage of meiosis, not completed in ovulation. With completion of meiosis, a second polar body is formed in the oocyte.

The haploid set from meiosis forms within the nuclear membrane and female pronucleus in the oocyte cell (Fig. 19.3 stage 3). After the spermatozoon has entered the oocyte and the haploid set entered the male pronucleus, the nuclear membrane breaks down, combining the two proneuclei together, forming a fertilized diploid set. A mitotic metaphase spindle is formed (Fig. 19.3 stage 4), then approximately 24 h later, mitosis occurs causing the first cell division with two identical cells generated.

MORULA AND BLASTOCYST FORMATION

The zygote then proceeds through multiple cellular divisions. While it took just one day to go from one to two identical cells, it takes two days to go from two identical cells to four identical cells, approximately 3 days then to make a 16 cell morula, and five days to go to a 32 cell blastocyst. During this time, the conceptus remains approximately the same size. Its size is limited by the constant size of the surrounding zona pellucida (Fig. 19.4). The result is that the cells become squeezed smaller and smaller to fit the space (Fig. 19.4). The conceptus cells produce multiple growth factors, insulin-like growth factor-1 and -2, transforming growth factors-α and β (TGF α and β), epidermal growth factor (EGF), and platelet-derived growth factor. All the growth factors are autocrines acting back on the conceptus cell to promote growth.

At around 8–16 cells, the conceptus undergoes its first differentiation, and forms a morula (Fig. 19.4). The first differentiation is into chorion or placental cells (Fig. 19.4, chorion cells shown in *yellow*) and amnion or fetal cells or "inner cell mass" (*red cells* in Fig. 19.4). The chorion cells are destined to become trophoblast placental cells and the amnion cells are destined to become the fetus. Approximately two days later, the morula becomes a blastocyst (Fig. 19.4). The blastocyst comprises placental cytotrophoblast cells and a more advanced inner cell mass. The cytotrophoblast cells surround a blastocoel filled with blastocoelic fluid. The formation of blastocysts and expansion of the blastocoelic cavity is promoted by EGF and TGFα.

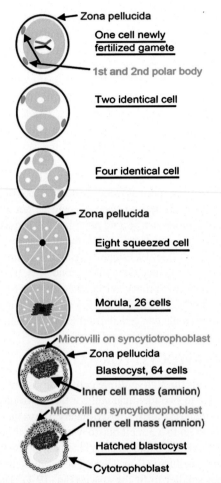

FIGURE 19.4 Growth of an embryo into a morula and a blastocyst.

While the blastocyst approaches the endometrial lining, the blastocyst protective barrier, the zona pellucida is shed or hatched (Fig. 19.4). There is evidence that signaling occurs, "We're coming, be ready," upon hatching, between the blastocyst and the endometrial wall, involving estradiol production and placental human chorionic gonadotropin (hCG) release by the blastocyst. Syncytiotrophoblast cells are formed and the blastocyst grows microvilli (Fig. 19.4). These microvilli are used for attachment of the blastocyst to the decidualized uterine epithelium.

TWINNING

The zona pellucida also functions in pregnancies to prevent the conceptus from coming apart. If the conceptus does break apart forming two distinct groups of cells or multiple groups of cells, monozygotic (coming from one zygote), twins

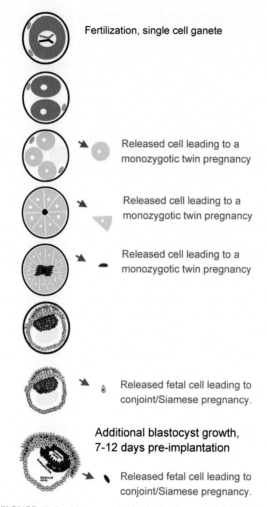

Fertilization, single cell ganete

Released cell leading to a
monozygotic twin pregnancy

Released cell leading to a
monozygotic twin pregnancy

Released cell leading to a
monozygotic twin pregnancy

Released fetal cell leading to
conjoint/Siamese pregnancy.

Additional blastocyst growth,
7-12 days pre-implantation

Released fetal cell leading to
conjoint/Siamese pregnancy.

FIGURE 19.5   Monozygotic identical twins and conjoint twins.

or multiple pregnancies result. These are genetically identical twins, triplets, or quadruplets. Monozygotic twins come from cells that have split prior to 12 days after fertilization (Fig. 19.5). After 12 days monozygotic twinning leads to conjoint and Siamese pregnancies (Fig. 19.5).

Di-, tri-, tetra-, penta-, hexa-, and hepta-zygotic twins result from independent spermatozoa fertilizing two to seven independent oocytes. Normally in women, just one oocyte is ovulated each menstrual cycle. Occasionally, two, and rarely three to seven oocytes are ovulated. Promotion of ovulation with hCG, most notably in IVF pregnancies, increases the chances of a multiple zygotic pregnancy and nonidentical twinning. The statistical odds of having a twin or multiple pregnancy are 1 in 90 pregnancies. The odds are 1 in 135 for nonidentical multiple zygotic pregnancies and approximately 1 in 270 for monozygotic identical twinning. The odds of conjoint twinning are 1 in 50,000, these may be Siamese twins, or twins that share certain organs.

FURTHER READING

[1] de Paula WBM, Lucas CH, Agip AA, Vizcay-Barrena G, Allen JF. Energy, ageing, fidelity and sex: oocyte mitochondrial DNA as a protected genetic template. Philos Trans Royal Soc Lond Biol Sci 2013;368:20120263.

[2] Taunton J, Rowning BA, Coughlin ML, Wu M, Moon RT, Mitchison TJ, et al. Actin-dependent propulsion of endosomes and lysosomes by recruitment of N-WASP. J Cell Biol 2000;148:519–30.

[3] Hoodbhoy T, Talbot P. Mammalian cortical granules: contents, fate and function. Mol Reprod Dev 1994;39:439–48.

[4] Kim NH, Funahashi H, Abeydeera LR, Moon SJ, Prather RS, Day BN. Effects of pig oocytes in vitro. J Reprod Fertil 1996;107:79–86.

[5] Miller DJ, Gong X, Deckar G, Shur BD. Egg cortical granule N-acetylglucosaminidase is required for the mouse zona block to polyspermy. J Cell Biol 1993;123:1431–40.

Chapter 20

Multiple Human Chorionic Gonadotropin Molecules

The hormone human chorionic gonadotropin (hCG) has always be considered as one molecule, this is in fact very wrong. Modern research shows that the hCG α- and β-subunit genes code for five or six completely independent molecules. The hormone hCG has always been claimed by text books on reproductive medicine and obstetrics and gynecology to have one principal function, maintaining progesterone production by corpus luteal cells. This is also very wrong. This is a secondary function of hCG. The principal function of hCG is maintenance of hemochorial placentation, progesterone; production in pregnancy is one of many secondary functions of the hormone hCG.

FIVE OR SIX COMPLETELY INDEPENDENT hCG MOLECULES

The hCG genes factually code for five of six completely independent molecules all sharing a common amino acid subunit sequence. First, there is the hormone hCG as produced in pregnancy, also call hCG-1. This is an α–β-subunit dimer of molecular weight 37,200 (Table 20.1). This molecule is produced by fused syncytiotrophoblast cells and has four biantennary N-linked oligosaccharides, structures MN, NN, and NNF (Fig. 20.1) and 4 Type 1 O-linked oligosaccharides structures N1 and N2 attached to the peptides (Fig. 20.1) [1–2].

The principal function of the hormone hCG or hCG-1 is in the production of hemochorial placentation which is completed at around 10 weeks of gestation [3–7] (Fig. 20.2). While hyperglycosylated hCG or hCG-2 promotes growth of placental cytotrophoblast cells [9,10], hCG promotes growth of uterine arteries so that they reach the hemochorial placentation structure [3,4] (Fig. 20.2), promotes fusion of cytotrophoblast cells to form syncytiotrophoblast cells [5], and promotes formation of the umbilical placenta–fetus circulation [6,7].

The hormone hCG has multiple secondary functions. These include promotion of uterine corpus luteal progesterone production in early pregnancy [8], suppression of uterine contraction during the course of pregnancy [9,10], and suppression of immune rejection and macrophage rejection of placental and fetal cells [11,12].

The second independent form of hCG is known at hyperglycosylated hCG, and is also called hCG-2. This is produced by cytotrophoblast cells in pregnancy. This molecule is an autocrine that acts on a transforming growth factor β

Biology of Life. http://dx.doi.org/10.1016/B978-0-12-809685-7.00020-4

TABLE 20.1 Multiple Semiindependent Variants of hCG. Molecular Weights (MW) Consider the Final Analysis of Elliott et al. [1], Valmu et al. [2] and the Recent Corrections of Cole [2]

Parameter	hCG	Hyperglycosylated hCG	Sulfated hCG	Cancer hCG	Hyperglycosylated hCG Free β-Subunit	Fetal hCG
New name	hCG-1	hCG-2	hCG-3	hCG-4	hCG-4 free β	hCG-5
Source cells	Syncytiotrophoblast	Cytotrophoblast	Gonadotrope	Trophoblastic malignancies	Nontrophoblastic malignancies	Fetal, kidney, and liver
Mode of action	Endocrine	Autocrine	Endocrine	Autocrine	Autocrine	ND
Total MW	37,200	39,800	36,150	41,280	27,220	Not determined
Site of action	LH–hCG receptor	TGFβ antagonism	LH–hCG receptor	TGFβ antagonism	TGFβ antagonism	Fetal organs
Amino acids α-subunit	92	92	92	92	–	ND
Amino acids β-subunit	145	145	145	145	145	ND
Peptide MW	26,200	26,200	26,200	26,200	16,000	ND
O-linked sugar units	4	4	4	4	4	ND
Type of O-linked sugars	Type 1	Type 2	Type 1+Sulfate	Type 2	Type 2	ND
N-linked sugar units	4	4	4	4	2	ND
Type of N-linked sugars	Biantennary	Biantennary	Biantennary	Triantennary β	Triantennary	ND
Sugar side chain MW	11,000	13,600	9,950	15,080	11,220	ND
Percentage sugars	30%	34%	28%	37%	41%	ND

hCG, human chorionic gonadotropin; LH, luteinizing hormone; ND, not determined; TGF, transforming growth factor.

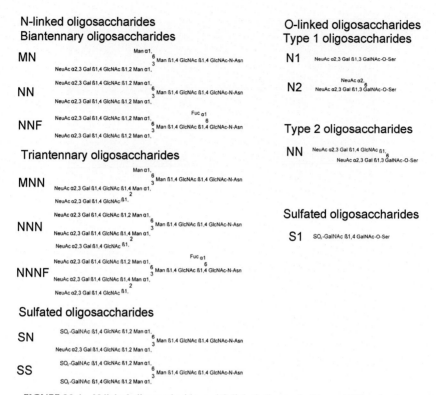

FIGURE 20.1 N-linked oligosaccharides and O-linked oligosaccharides on hCG molecules.

(TGFβ) receptor rather that the hormone's hCG/LH receptor [13,14]. It has a larger molecular weight of 39,800 than hCG-1, and contains four biantennary N-linked oligosaccharides structures MN, NN and NNF (Fig. 20.1), and four unique Type 2 O-linked oligosaccharides NN [1,2] (Fig. 20.1). Somehow, the Type 2 O-linked oligosaccharides make this a very independent molecule, an autocrine rather than a hormone [15].

Hyperglycosylated hCG, hCG-2, has completely independent functions to hCG-1. It controls invasion and all aspects of implantation of the blastocyst in pregnancy [15,16]. It also controls placental growth during the course of pregnancy [17,18]. It controls growth of cytotrophoblast cells, these cells continuously fuse to make syncytiotrophoblast cells during the course of pregnancy.

The third independent type of hCG produced is pituitary sulfated hCG or hCG-3. hCG evolved from luteinizing hormone (LH) in lower simian primates [19]. As such, it is no surprise that hCG-3 is a pituitary gonadotrope cell hormone like LH. hCG-3 is smaller than hCG-1 with a molecular weight of 36,150 verses 37,200 (Table 20.1). It has rather different sulfated oligosaccharides, having a mixture of structures NN, SN, and SS N-linked oligosaccharides and S1, N1, and N2 O-linked oligosaccharides (Fig. 20.1).

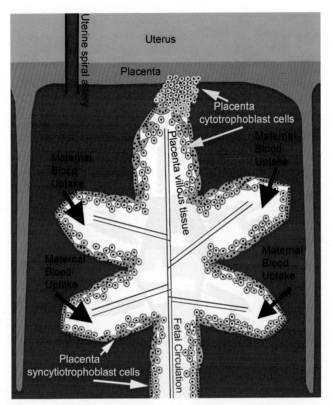

FIGURE 20.2 Hemochorial placentation. The placenta contains four to six chambers.

hCG-3 acts like LH on a common LH–hCG hormone receptor, involved in estradiol synthesis in the follicular phase of the menstrual cycle, and progesterone synthesis in the luteal phase of the menstrual cycle. It similarly supplements LH in testosterone synthesis in men. In women it works with LH in controlling ovulation. It is thought that approximately 50% of the biological activity in these processes comes from LH (high concentration, low potency) and 50% from hCG (low concentration, high potency) [20].

hCG and it free β-subunit are also produced in variable concentrations by most human malignancies. Trophoblastic malignancies, choriocarcinoma, testicular germ cell malignancies, and ovarian germ cell malignancies produce hCG α−β-subunit dimer, a superglycosylated hCG or hCG-4. This is hCG with triantennary N-linked oligosaccharides structures MNN, NNN, and NNNF (Fig. 20.1), and Type 2 O-linked oligosaccharides structure NN (Fig. 20.1). All other malignancies or nontrophoblastic malignancies produce an hCG-4 free β. The two molecules are the largest forms of hCG and its free β-subunit produced, molecular weight 41,280 and 27,220 (Table 20.1).

Research now shows that hCG-4 and hCG-4 free β act on a TGFβ receptor driving cancer cells [13,21], promoting cancer cell growth, blocking cancer cell apoptosis, and promoting cancer cell production of proteases or cancer cell invasion [21,22].

Other research shows that a variant of hCG is made by fetal kidney and fetal liver during pregnancy. This promotes growth of fetal organs [23,24]. We do not know if this is an established variant of hCG like hCG-1, hCG-2, hCG-3, or hCG-4 or a separate variant because structure and molecular weight studies have not been performed [23,24]. As such, we say that there is five or six independent variants of hCG.

REFERENCES

[1] Elliott MM, Kardana A, Lustbader JW, Cole LA. Carbohydrate and peptide structure of the alpha- and beta-subunits of human chorionic gonadotropin from normal and aberrant pregnancy and choriocarcinoma. Endocrine 1997;7:15–32.

[2] Valmu L, Alfthan H, Hotakainen K, Birken S, Stenman UH. Site-specific glycan analysis of human chorionic gonadotropin β-subunit from malignancies and pregnancy by liquid chromatography - electrospray mass spectrometry. Glycobiology 2006;16:1207–18.

[3] Toth P, Li X, Rao CV, Lincoln SR, Sanfillipino JS, Spinnato JA, et al. Expression of functional human chorionic gonadotropin/human luteinizing hormone receptor gene in human uterine arteries. J Clin Endocrinol Metab 1994;79:307–15.

[4] Zygmunt M, Herr F, Keller-Schoenwetter S, Kunzi-Rapp K, Munstedt K, Rao CV, et al. Characterization of human chorionic gonadotropin as a novel angiogenic factor. J Clin Endocrinol Metab 2002;87:5290–6.

[5] Shi QJ, Lei ZM, Rao CV, Lin J. Novel role of human chorionic gonadotropin in differentiation of human cytotrophoblasts. Endocrinology 1993;132:387–95.

[6] Rao CV, Li X, Toth P, Lei ZM, Cok VD. Novel expression of functional human chorionic gonadotropin/luteinizing hormone receptor in human umbilical cords. J Clin Endocrinol Metab 1993;77:1706–14.

[7] McGregor WG, Raymoure WJ, Kuhn RW, Jaffe RB. Fetal tissues can synthesize a placental hormone evidence for chorionic gonadotropin β-subunit synthesis by human fetal kidney. J Clin Invest 1981;68:306–9.

[8] Strott CA, Yoshimi T, Ross GT, Lipsett MB. Ovarian physiology: relationship between plasma LH and steroidogenesis by the follicle and corpus luteum; effect of HCG. J Clin Endocrinol Metab 1969;29:1157–67.

[9] Eta E, Ambrus G, Rao V. Direct regulation of human myometrial contractions by human chorionic gonadotropin. J Clin Endocrinol Metab 1994;79:1582–6.

[10] Doheny HC, Houlihan DD, Ravikumar N, Smith TJ, Morrison JJ. Human chorionic gonadotrophin relaxation of human pregnant myometrium and activation of the BKCa channel. J Clin Endocrinol Metab 2003;88:4310–5.

[11] Akoum A, Metz CN, Morin M. Marked increase in macrophage migration inhibitory factor synthesis and secretion in human endometrial cells in response to human chorionic gonadotropin hormone. J Clin Endocrinol Metab 2005;90:2904–10.

[12] Wan H, Marjan A, Cheung VW, Leenen PJM, Khan NA, Benner R, et al. Chorionic gonadotropin can enhance innate immunity by stimulating macrophage function. J Leukoc Biol 2007;82:926–33.

[13] Butler SA, Ikram MS, Mathieu S, Iles RK. The increase in bladder carcinoma cell population induced by the free beta subunit of hCG is a result of an anti-apoptosis effect and not cell proliferation. Br J Cancer 2000;82:1553–6.

[14] Berndt S, Blacher S, Munuat C, Detilleux J, Evain-Brion D, Noel A, et al. Hyperglycosylated human chorionic gonadotropin stimulates angiogenesis through TGF-ß receptor activation. J Fed Am Soc Exp Biol 2013;27:1309–21.

[15] Sasaki Y, Ladner DG, Cole LA. Hyperglycosylated hCG the source of pregnancy failures. Fertil Steril 2008;89:1786–871.

[16] Cole LA. Hyperglycosylated hCG and pregnancy failures. J Reprod Immunol 2012;93:119–22.

[17] Cole LA, Dai D, Butler SA, Leslie KK, Kohorn EI. Gestational trophoblastic diseases: 1. Pathophysiology of hyperglycosylated hCG-regulated neoplasia. Gynecol Oncol 2006;102:144–9.

[18] Brennan MC, Wolfe MD, Murray-Krezan CM, Cole LA, Rayburn WF. First trimester hyperglycosylated human chorionic gonadotropin and development of hypertension. Prenat Diagn 2013;33:1075–9.

[19] Fiddes JC, Goodman HM. The gene encoding the common alpha subunit of the four glycoprotein hormones. J Mol Appl Genet 1981;1:3–18.

[20] Cole LA, Gutierrez JM. Production of hCG during the menstrual cycle. J Reprod Med 2009;54:245–50.

[21] Cole LA, Butler SA. Hyperglycosylated hCG, hCGß and hyperglycosylated hCGß: interchangeable cancer promoters. Mol Cell Endocrinol 2012;349:232–8.

[22] Cole LA, Butler SA. B152 anti-hyperglycosylated human chorionic gonadotropin free β-Subunit. A new, possible treatment for cancer. J Reprod Med 2015;60:13–20.

[23] Goldsmith PC, McGregor WG, Raymoure WJ, Kuhn RW, Jaffe RB. Cellular localization of chorionic gonadotropin in human fetal kidney and liver. J Clin Endocrinol Metab 1983;57:54–61.

[24] Huhtaniemi IT, Korenbrot CC, Jaffe RB. Content of chorionic gonadotropin in human fetal tissue. J Clin Endocrinol Metab 1978;46:994–9.

Chapter 21

Implantation and Pregnancy Failure

Research strongly indicates that hyperglycosylated human chorionic gonadotrophin (hCG), hCG-2, drives pregnancy implantation. Hyperglycosylated hCG drives cytotrophoblast cell invasion and growth [1]. Inefficient implantation cause miscarriages of pregnancies, biochemical pregnancies, and ectopic pregnancies [2–5]. Inefficient implantation is due to an inefficient supply of hyperglycosylated hCG [2–5]. Antibodies to hyperglycosylated hCG blocks implantation. From the findings it is inferred that hyperglycosylated hCG drives implantation. As demonstrated, hyperglycosylated hCG binds and antagonizes a transforming growth factor (TGF)β receptor [6,7]. As shown, this antagonization of TGFβ promotes metalloproteinases and collagenase production, leading to physical invasion as occurs in implantation [8,9].

IMPLANTATION

Syncytiotrophoblast cells on a blastocyst develop microvilli. The blastocyst hatches (Fig. 21.1, Panel A). With hatching, the syncytiotrophoblast cells produce hCG, estradiol, and EGF which signal to the uterine epithelium that a blastocyst is about to implant (Fig. 21.1, Panel B) [8,9].

The microvilli connect with the pinopodes on the uterine epithelium linking the blastocyst to the uterus (Fig. 21.2, Panel C). The cytotrophoblast cells on the blastocyst produce hyperglycosylated hCG which drives cytotrophoblast cell invasion of the uterus (Fig. 21.2, Panel C). The blastocyst enters the uterus (Fig. 21.2, Panel D). The hyperglycosylated hCG driven invasion continues and the uterine epithelium seals itself closed burying the cytotrophoblast inside the uterus (Fig. 21.3, Panel E).

The hyperglycosylated hCG invasion continues driving invasion to >30% of thickness of the uterus (Fig. 21.3, Panel F). As hyperglycosylated hCG drives the invasion of the blastocyst, it also drives growth of cytotrophoblast cells in the blastocyst complex. Blastocyst columns of cytotrophoblast cells develop (Fig. 21.3, Panel E), these grow into villous structures of hemochorial placentation (Fig. 21.3, Panel F).

Biology of Life. http://dx.doi.org/10.1016/B978-0-12-809685-7.00021-6

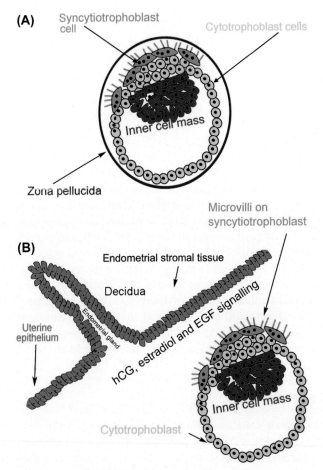

FIGURE 21.1 Human implantation part 1, approach of uterine epithelium.

PREGNANCY FAILURE

There are three clear types of pregnancy failure. The first type are miscarriages or spontaneous abortions, which occur in the first and second trimester of pregnancy; they account for 17% of pregnancies. The second type are biochemical pregnancies, which occur in the fifth to sixth week of gestation; these account for 25% of pregnancies. Finally, there are ectopic or tubal pregnancies; they account for 2% of all pregnancies. As found by Norwitz et al. [10], Semprini and Simon [11], Walker et al. [12], and Wilcox et al. [13], miscarriage and biochemical pregnancies are mostly due to improper implantation.

Here we present published results on 298 pregnancies [5]. A total of 191 pregnancies ended with a normal term delivery, 49 ended in miscarriage or spontaneous abortion, 40 ended in biochemical pregnancy, and 18 ended in

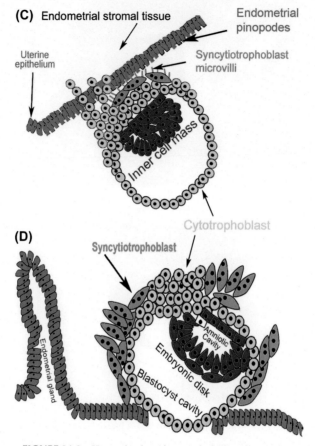

FIGURE 21.2   Human implantation part 2, the invasion process.

ectopic pregnancy (Fig. 21.4). Of the normal term birth pregnancies, 191 of 191 (or 100%) produced greater than 40% hyperglycosylated hCG. In contrast, 36 of 49 spontaneous abortion/miscarriage pregnancies, 31 of 40 biochemical pregnancies and 18 of 18 ectopic pregnancies produced less than 40% hyperglycosylated hCG (Fig. 21.4). This indicates that pregnancy failure is mostly due to insufficient supply of hyperglycosylated hCG.

It is interesting that 18 of 18 ectopic pregnancies were associated with deficient hyperglycosylated hCG. It is possible that only pregnancies deficient in hyperglycosylated hCG can implant ectopically. The fallopian tube has a thin wall. It is possible that normal pregnancies would invade their way through the full thickness of the fallopian tube and enter inter-tissue space. As such, all ectopic pregnancies are those that would fail due to improper implantation. It appears that no normal pregnancies become ectopic pregnancies. It is inferred that most failing pregnancies are due to insufficient generation of hyperglycosylated hCG.

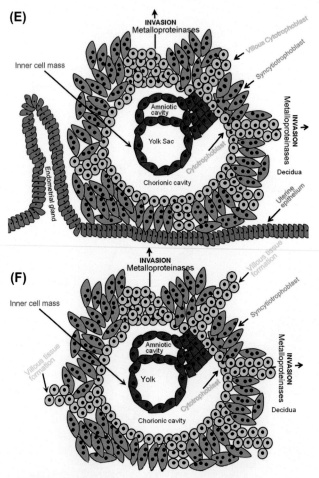

FIGURE 21.3 Human implantation part 3, burial deep into the uterus.

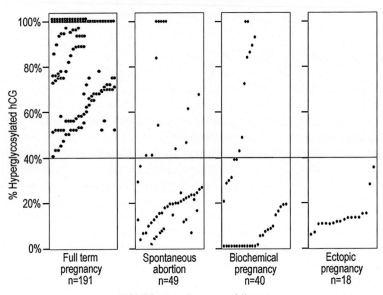

FIGURE 21.4 Pregnancy failures.

REFERENCES

[1] Cole LA, Dai D, Butler SA, Leslie KK, Kohorn EI. Gestational trophoblastic diseases: 1. Patho-physiology of hyperglycosylated hCG-regulated neoplasia. Gynecol Oncol 2006;102:144–9.

[2] Sasaki Y, Ladner DG, Cole LA. Hyperglycosylated hCG the source of pregnancy failures. Fertil Steril 2008;89:1781–6.

[3] Cole LA. Hyperglycosylated hCG and pregnancy failures. J Reprod Immunol 2012;93:119–22.

[4] Cole LA, Khanlian SA, Kohorn EI. Evolution of the human brain, chorionic gonadotropin and hemochorial implantation of the placenta: origins of pregnancy failures, preeclampsia and choriocarcinoma. J Reprod Med 2008;53:449–557.

[5] Cole LA. Chapter 19: placentation disorders and failures of early pregnancy. In: Cole LA, editor. The biology and medical dynamics of human reproduction. New York: Nova Science Publishing; 2013.

[6] Butler SA, Ikram MS, Mathieu S, Iles RK. The increase in bladder carcinoma cell population induced by the free beta subunit of hCG is a result of an anti-apoptosis effect and not cell proliferation. Brit J Cancer 2000;82:1553–6.

[7] Berndt S, Blacher S, Munuat C, Detilleux J, Evain-Brion D, Noel A, et al. Hyperglycosylated human chorionic gonadotropin stimulates angiogenesis through TGF-ß receptor activation. J Fed Am Soc Exp Biol 2013;27:1309–21.

[8] Ohlsson R, Larsson E, Nilsson O, Wahlstrom T, Sundstrom P. Blastocyst implantation pre-cedes induction of insulin-like growth factor II gene expression in human trophoblasts. Devel-opment 1989;106:555–9.

[9] d'Hauterivea SP, Berndt BS, Tsampalasa M, Charlet-Renarda C, Dubois M, Bourgain C, et al. Dialogue between blastocyst hCG and endometrial hCG/LH receptor: which role in implanta-tion? Gyncol Obstet Invest 2007;64:156–60.

[10] Norwitz ER, Schust DJ, Fisher SJ. Implantation and the survival of early pregnancy. N Engl J Med 2001;345:1400–8.

[11] Semprini AE, Simon G. Not so efficient reproduction. Lancet 2000;56:257–358.

[12] Walker EM, Lewis M, Cooper W, Marnie M, Howie PW. Occult biochemical pregnancy: fact or fiction. Br J Obstet Gynaecol 1988;95:659–63.

[13] Wilcox AJ, Weinberg CR, Baird DD. Risk factors for early pregnancy loss. Epidemiol 1990;1:382–5.

Chapter 22

Hemochorial Placentation

Fetal life is dependent on unfiltered maternal uterine blood supply until the development of an independent blood supply, by hemochorial placentation. It is appropriate to say that the fetus is like the development of an extra organ within the mother until hemochorial placentation. Hemochorial placentation is the time that the fetus has an independent circulation making it an independent being. It is reasonable to say that the fetus does not become an independent human being until hemochorial placentation. Yes, with the development of hemochorial placentation the fetus is still not totally independent, as it is still dependent on the mother for oxygen, sugar, some amino acids, and basic nutrient supply. In addition, it is not until birth that the fetus can take a breath of fresh air, and becomes a fully independent human being.

Hemochorial placentation is the name for advanced placental function, whereby the placenta separates the maternal and fetal circulation, and the placenta transfers oxygen, sugars, and essential nutrients to the independent fetal circulation. Research indicates that hemochorial placentation is instituted by the action of the hormone human chorionic gonadotrophin (hCG) and the autocrine hyperglycosylated hCG.

Fig. 22.1 indicates that both the hormone hCG and the autocrine hyperglycosylated hCG are produced throughout pregnancy. Hyperglycosylated hCG has been shown to promote cytotrophoblast tissue and villous cytotrophoblast tissue growth and invasion throughout pregnancy [1–5]. It is the action of this autocrine that promotes formation of villous cytotrophoblast columns (Fig. 22.2) and the development of cytotrophoblast columns (Fig. 22.3) during pregnancy. These cytotrophoblast columns form tree-like structures or completed villi (Fig. 22.4).

As cytotrophoblast villi are completed, driven by the autocrine hyperglycosylated hCG, the hormone hCG promotes fusion of cytotrophoblast cells to form a skin of syncytiotrophoblast tissue that surrounds the cytotrophoblast villous structures [6] (Figs. 22.3 and 22.4). This completed the villous structures.

The hormone hCG promotes expansion of the spiral uterine arteries to reach the villous placental structure [7,8] (Fig. 22.5), and formation of the umbilical circulation to link the hemochorial placentation and the fetal circulation [9,10] (Fig. 22.5).

Biology of Life. http://dx.doi.org/10.1016/B978-0-12-809685-7.00022-8

163

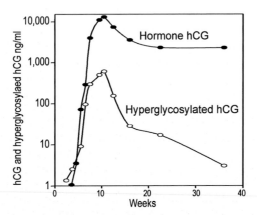

FIGURE 22.1 hCG and hyperglycosylated hCG serum concentrations during pregnancy.

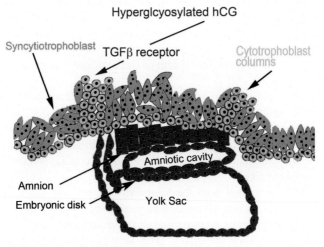

FIGURE 22.2 Cytotrophoblast columns, first stage of villous tissue formation.

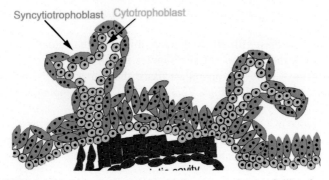

FIGURE 22.3 Extension of cytotrophoblast columns, second stage of villous formation.

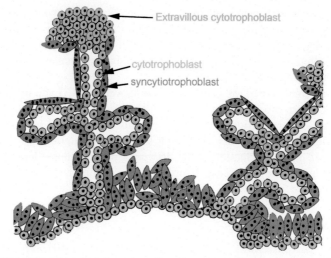

FIGURE 22.4   Formation of complete villous structure, third stage of villous formation.

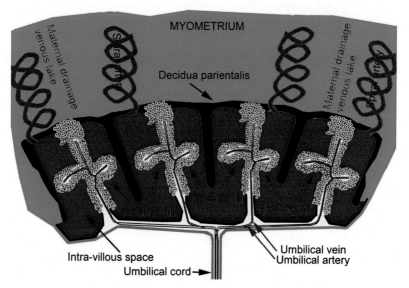

FIGURE 22.5   Hemochorial placentation.

Collectively, hCG and hyperglycosylated hCG drive hemochorial placentation. Hemochorial placentation is not completed until 10 weeks of gestation. Up until this time, fetal and placental tissues are supported by the maternal circulation, like a new organ linked to the maternal uterine circulation.

In active hemochorial placentation (Fig. 22.5), the villous unit is stuck to the maternal uterus myometrium muscle through a plug of cytotrophoblast cells called the extravillous cytotrophoblast cells (Fig. 22.4). In hemochorial placentation, the fetus has a separate blood circulation to the mother. Nutrients like oxygen and glucose pass through a single layer of syncytiotrophoblast tissue to enter the villous unit, they then enter the fetal umbilical circulation to be transported to the fetus (Fig. 22.5).

For the remainder of the pregnancy, 10–40 weeks, hCG and hyperglycosylated hCG are produced and expand upon villous cytotrophoblast tissue growth and differentiation with growing villous tissue and growing fetus with the length of pregnancy. Research shows that a deficiency in hyperglycosylated hCG production at 10–40 weeks of pregnancy, leads to preeclampsia and other hypertense disorders in the second and third trimesters of pregnancy [5,11].

REFERENCES

[1] Cole LA, Dai D, Butler SA, Leslie KK, Kohorn EI. Gestational trophoblastic diseases: 1. Pathophysiology of hyperglycosylated hCG-regulated neoplasia. Gynecol Oncol 2006;102:144–9.

[2] Cole LA, Khanlian SA, Riley JM, Butler SA. Hyperglycosylated hCG (hCG-H) in gestational implantation and in choriocarcinoma and testicular germ cell malignancy tumorigenesis. J Reprod Med 2006;51:919–29.

[3] Handschuh K, Guibourdenche J, Tsatsari V, Guesnon M, Laurendeau I, Evain-Brion D, et al. Human chorionic gonadotropin produced by the invasive trophoblast but not the villous trophoblast promotes cell invasion and is down-regulated by peroxisome proliferator-activated receptor-gamma. Endocrinology 2007;148:5011–9.

[4] Guibourdenche J, Handschuh K, Tsatsaris V, Gerbaud MC, Legul F, Muller D, et al. Hyperglycosylated hCG is a marker of early human trophoblast invasion. J Clin Endocrinol Metab 2010;95:E240–4.

[5] Brennan MC, Wolfe MD, Murray-Krezan CM, Cole LA, Rayburn WF. First trimester hyperglycosylated human chorionic gonadotropin and development of hypertension. Prenat Diagn 2013;33:1075–9.

[6] Shi QJ, Lei ZM, Rao CV, Lin J. Novel role of human chorionic gonadotropin in differentiation of human cytotrophoblasts. Endocrinology 1993;132:387–95.

[7] Toth P, Li X, Rao CV, Lincoln SR, Sanfillipino JS, Spinnato JA, et al. Expression of functional human chorionic gonadotropin/human luteinizing hormone receptor gene in human uterine arteries. J Clin Endocrinol Metab 1994;79:307–15.

[8] Zygmunt M, Herr F, Keller-Schoenwetter S, Kunzi-Rapp K, Munstedt K, Rao CV, et al. Characterization of human chorionic gonadotropin as a novel angiogenic factor. J Clin Endocrinol Metab 2002;87:5290–6.

[9] Rao CV, Li X, Toth P, Lei ZM, Cok VD. Novel expression of functional human chorionic gonadotropin/luteinizing hormone receptor in human umbilical cords. J Clin Endocrinol Metab 1993;77:1706–14.

[10] Rao CV, Li X, Toth P, Lei ZM. Expression of epidermal growth factor transforming growth factor-alpha and their common receptor genes in human umbilical cords. J Clin Endocrinol Metab 1995;80:1012–20.

[11] Bahado-Singh RO, Oz AU, Kingston JM, Shahabi S, Hsu CD, Cole LA. The role of hyperglycosylated hCG in trophoblast invasion and the prediction of subsequent pre-eclampsia. Prenat Diagn 2002;22:478–81.

Chapter 23

Human Life, Development of the Human Brain

It is the brain that primarily distinguishes all humans from other species. The extremely large human brain is 2.4% of total body weight at birth (Table 23.1). Other species have birth brains which are 0.07–0.8% of total body weight. It is the developed human brain, then thinking, talking, planning, plotting, and conceiving that makes a human a human, and distinguish humans from all other species. The human brain is developed in the fetus. I ask the political questions when does the fetus officially develop the human brain and when does the fetus really become human with a large 2.4% brain?

The brain starts in prenatal or fetal life, just three weeks after conception. It continues to develop throughout fetal life, but in many ways its development is a lifelong project. This is because the actions that shape the brain during development are also accountable for storing information, new skills, and memories, throughout life. The major difference between brain advancement in a child versus learning in an adult is a matter of degree; the brain is far more impressionable in a child.

As shown in Fig. 23.1 the fetus continuously grows in size during pregnancy. As shown in Fig. 23.2 the brain continuously grows in size during pregnancy. As shown in Fig. 23.3, heart development, eyes development, leg, arms, palate, genitalia, and ear development can be attributed to specific periods in fetal life but brain and central nervous system development is different, it is continuous 3–40 weeks of gestation or through the whole length of gestation. Clearly, examining Table 23.2, the greatest amount of growth occurs in the third trimester or 26–40 weeks of gestation.

Looking at fetal structure you can say that the fetus does not look human until approximately 16 weeks of gestation (Fig. 23.2). But when brain development wise does the fetus become fetal development wise truly human? Looking at brain growth and body weight during pregnancy (Table 23.2), at two months or eight weeks gestation the fetus has about 30,000 neuron cells and weighs 1 g or 30,000 cells/g; at three months or 13 weeks the fetus has 1,000,000 neuron cells and weighs around 28 g or 43,000 cells/g; at six months or 26 weeks the fetus has 250 million neuron cells and weighs around 760 g or 330,000 cells/g; and at term or 40 weeks the fetus has about 10,000 million neuron cells and

Biology of Life. http://dx.doi.org/10.1016/B978-0-12-809685-7.00023-X

TABLE 23.1 Brain Size

Species	Brain Size (% of Total Body Weight)
Homo sapiens	2.4
Dog	0.80
Cat	0.78
Chimpanzee	0.74
Lion	0.18
Elephant	0.18
Horse	0.17
Kangaroo	0.16
Giraffe	0.13
Camel	0.13
Whale	0.1
Cow	0.09
Pig	0.09
Prosimian primate	0.07

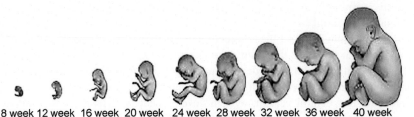

8 week 12 week 16 week 20 week 24 week 28 week 32 week 36 week 40 week

FIGURE 23.1 Fetus development during pregnancy. *Figure prepared from illustration* www.calculatorbirds.com/pregnancy/information-about-fetal-development-during-pregnancy.html.

weighs around 3400 g or 2,941,000 cells/g [1,2] (Table 23.2). Clearly the fetus is truly not human with a large 2.4% brain until 40 weeks when the fetus is born.

The fetus has at least seven genes that promote exponential brain growth during fetal life, these are the MCPH1, ASPM, CD5RAP2, CENPJ, WDR62, CEP152, and STIL genes. Each promotes exponential growth in brain size [1,2].

Clearly, brain growth increases disproportionately to body weight with advancing pregnancy, expanding the most during the last three months of pregnancy. There is an exponential increase in the brain size with advancing fetal weight. So, when you consider that it is the large brain size that

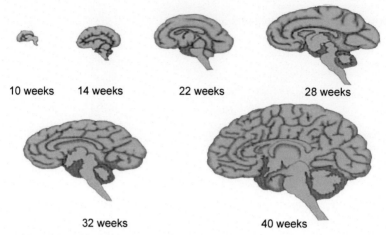

FIGURE 23.2 Brain development during pregnancy. *Figure prepared from illustration* www.beginbeforebirth.org/in-the-womb/fetal-pregnancy.

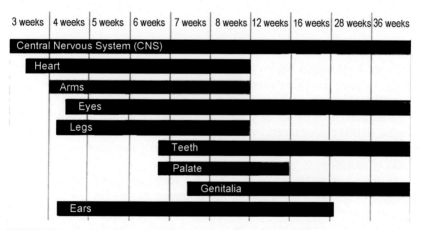

FIGURE 23.3 Brain development during pregnancy. *Figure prepared from National Institutes of Alcohol Abuse and Alcoholism illustration, Module 10k Fetal Alcohol Exposure.*

distinguishes a human from other species , you cannot really call a fetus a human until 32–40 weeks of gestation or until the fetus is fully developed and delivered.

Considering that term pregnancy has a brain weight of 2.4% of body weight, and comprises 10,000 million cells, then brain weight at six months pregnancy comprises just 0.27% of body weight, brain weight at three months of pregnancy comprises just 0.029% of body weight, and brain weight at two months or eight weeks of pregnancy comprises just 0.00026% of body weight (Table 23.2). At two months and three months pregnancy brain weight is smaller than any known animal species (Table 23.1).

TABLE 23.2 Fetal Weight and Brain Size [1–3]

Gestational Age	Fetal Weight (g)	No. of Neurons (Cells)	No. of Neurons/ Fetal Weight (Cells/g)	Brain/ Body Weight (%)
8 weeks (2 months)	1	30,000	30,000	0.00026
13 weeks (3 months)	28	1,000,000	43,000	0.029
26 weeks (6 months)	760	250,000,000	330,000	0.27
40 weeks (9 months)	3400	10,000,000,000	2,941,000	2.4

Fetal weight from http://www.babycentre.co.uk/a1004000/average-fetal-length-and-weight-chart and number of neurons from http://www.hhmi.org/biointeractive/development-human-embryonic-brain.

Brain development begins with the closure of the neural tube at around 27 days or three weeks six days of gestation (assuming three weeks implantation). With closure, the tube transforms into tissues destined to be the brain and tissue destined to be the spinal cord or the embryo. At around five weeks, there may be 10,000 neuronal cell. At five weeks of gestation, the first synapses are formed in the spinal cord. This permits the first fetal movement at around six weeks of gestation.

At around eight weeks of gestation the fetus can move legs, at around 10 weeks of gestation the fetus can move fingers. At 18 weeks of gestation a mother can feel movements made by the fetus. At around 14 weeks of gestation a fetus makes rhythmic contractions of the diaphragm, coordinated sucking and swallowing reflexes, or abilities controlled by the brain stem.

At 32–40 weeks of gestation the cerebral cortex develops mental life. The fetus starts thinking, remembering, and feeling. Premature babies show basic cerebral cortex activities including the capability of learning. Responding to familiar odors such as amniotic fluid, sound such as mother's heartbeat or mother's voice, and sensory qualities of the womb.

In conclusion, the fetus seemingly evolves to become a species more and more resembling human as pregnancy progresses, becoming, in regards of brain size, a human at term. The early fetus with a miniscule brain, at eight weeks gestation, visually resembles a shrimp (Fig. 23.1); at this time, the brain is just 0.00026% of body weight and seemingly, is no more mentally or neurologically advanced than a shrimp. This early pregnancy state cannot be considered as human.

REFERENCES

[1] Average fetal length and weight chart. http://www.babycentre.co.uk/a1004000/average-fetal-length-and-weight-chart.

[2] Number of neurons in fetus. http://www.hhmi.org/biointeractive/development-human-embryonic-brain.

[3] Child development in 3rd trimester. http://mom.me/pregnancy/12467-pregnancy-brain-and-child-development-third-trimester/.

FURTHER READING

[1] Dalby JT. Environmental effects on prenatal development. J Pediatr Psychol 1978;3:105–9.

[2] Becher JC. Insights into early fetal development. Royal College of Physicians of Edinburg; 2004.

[3] Glenn OA. Normal development of the fetal brain by MRI. Semin Perinatol 2009;33:208–19.

[4] Monteagudo A, Timor-Tritsch IE. Ultrasound of the fetal brain. Ultrasound Clin 2007;2:217–44.

Chapter 24

Human Life

These Section V chapters have carefully looked at human development. Starting with semen and with the oocyte, these chapters have carefully examined human development. If you can define a human as an intellectual being that differs from all other species by its large brain (2.4% of body weight), then a human fetus is clearly not human-like with a big brain (2.4% of body weight) until it is fully developed at term pregnancy (Chapter 23).

I know that in the state of Texas and several other states, human life in the context of law and abortion is legally considered to start at fertilization. Examining basic science as presented here (Chapters 17–23), life at the time of fertilization is certainly not human or human-like, and the fertilized human zygote is actually a dying being with no source of nutrition or nutrients until it becomes implanted in the uterus 7 day after fertilization. It is estimated that the majority of fertilization are destroyed beings, in that they never get implanted into the uterus (see Chapter 21). How can the state of Texas then ever define a fertilized oocyte as a human being. As discussed in Chapter 23, the only pregnancy that can be truly considered as human-like is at 32–40 weeks.

In 2004 I had a strange run in with the state of Texas regarding an early embryo, 4–5 weeks gestation being a human being. If you check Laurence Cole Ph.D. out on the internet you will find out that he is a world expert on hCG and human pregnancy testing. He has over 390 peer reviewed medical articles on hCG and pregnancy testing. In 2004 I was phoned by the police in Galveston, TX.

As they related, they needed my help with pregnancy testing. All they had was stomach fluid from a young girl who was deceased. As they related, someone is accused of her murder, if she was pregnant when murdered it would be under Texas law, two murders, which was subject under Texas law to the death penalty (someone would be executed because they killed a 4–5 week fetus). I mistakenly said yes of course, I would try to detect pregnancy.

I related everything to my senior technician at that time, Sarah Khanlian, and we both found ourselves digging up information on the murder case on the internet. Apparently, a boy from the poor neighborhood of Galveston was having a relationship with a young girl from the very fancy neighborhood of Galveston. The parents condemned her relationship with this rough poor-side of town boy. The girl's parents had just left town for three weeks. The boy visited the girl at her house. When the boy visited, the girl was crying. She had just been to see her doctor who told her that she was four or five weeks pregnant. As

Biology of Life. http://dx.doi.org/10.1016/B978-0-12-809685-7.00024-1

the girl kept saying, how can I ever tell my parents when they return that I am pregnant by you, as you know they kept telling me to stay away from you. The boy responded, "don't worry you can have one of those early abortions." The girl responded "my parents will never allow that." The boy later left and went home.

The next day the boy came over to the girl's home again. He knocked on the front door of the house and nobody replied. He noticed that the door was not locked. He entered, the girl was not downstairs and so he went upstairs to her room. When he entered the girl's room he noticed that girl was in bed. That she had made a noose and tied it to the ceiling beams. The girl was trying to force her head into the noose. "What are you doing?," the boy shouted. "I cannot stand this pregnancy and having to deal with this with my parents," the girl responded. "Don't be silly, stop what you are doing" said the boy, "we will deal with this and you will get a quick and quiet abortion." "My parents won't allow this," said the girl, "here can you help me get this rope that I am struggling with over my chin." The boy helped her push the rope over her chin. The boy said "stop this, we will get you an abortion or come and lived with me." The girl quickly hung herself. There was nothing that the boy could do.

The girl's parents came home two weeks later and to their shock found their daughter's dead body hung in her bedroom. She was very briefly examined by a close friend of the parents, who was a pathologist and a state coroner, who did not detect the very early pregnancy. The coroner signed the death certificate. The only thing kept was a sample of her stomach fluids. The girl was very quickly cremated, and the parents learned nothing about what really happened.

One week later the boy went to see the movie "The Passion of Christ." He came out of the movie all loving of Christ, and saying to himself "Christ would want me to tell the police and the parents what exactly happened with that girl and why she hung herself." That evening he went to Galveston police without talking to a lawyer or anybody else, and related the whole factual story of exactly what had happened.

It is my personal opinion that the police should have given him a single spank on the bottom and sent him home. That is not what happened. Texas has no assisted suicide law, as such, the police immediately arrested the boy for murder for getting the tight noose over the girls chin. The police wanted to find the boy guilty of two murders, the girl and her 4–5 week baby. They could then demand, in court, his execution, but first they needed to prove that the girl was pregnant. This is where I entered the story and wrongly agreed to test her stomach fluids.

hCG is the hormone produced by the placenta that identifies pregnancy. I estimate from the fact that the doctor said that the girl was 4–5 weeks pregnant that she was approximately one or two weeks after pregnancy implantation with quite significant hCG concentration. The morning following the phone call with Galveston police the frozen sample of stomach fluid arrived by Federal Express. Yes, a small amount of hCG would enter her stomach fluids and be broken down by the stomach acids. I had a test for β-core fragment+free β-subunit+nicked free β-subunit, the terminal digestion or breakdown products of hCG. To honestly perform a test this is what I would need to measure. I ran the test for these molecules myself, everybody else in my laboratory refused, and true enough, the test was positive for hCG breakdown products or for pregnancy.

I said to myself "how can I report this test result to Galveston police? The boy will be executed in Texas for a double murder. This is very wrong. That boy does not deserve any of this, his crime is nonexistent." I made an appointment to see an Albuquerque lawyer. Following the lawyer's advice I wrote a letter to Galveston police saying that I performed a test for hCG fragments, and that the test was positive, given the actual test results. That does not, however, mean that the girl was definitively pregnant. An hCG test is not designed to work on stomach fluids and in all likelihood the acid and other component of stomach fluid interfered with the test and it is was possibly falsely positive.

On this basis, Galveston police were never able to charge the boy with a double murder. The boy is currently in prison in Texas for 12 long years for a single murder, and will be released shortly. This girl was just 4–5 weeks pregnant, considering the text of the preceding chapter this is like she has something much tinier than a tiny shrimp inside her, something that looks more like a shrimp than a human, something without a measurable brain. How can this ever be really considered as a human, and how can it's death ever be correctly considered as a murder, as it can be under Texas law?

This article clearly examines human life and human conception in detail. Clearly human life is distinguished from multiple animal lives by thinking, by brain function, and brain size. Under these definitions, fetal lives really cannot be considered as human-like until 32–40 weeks of pregnancy. Early human fetal life is very similar or parallel to primitive animal life.

Abortion experts all of the United States never seen to stop talking about "When can you terminate a fetus?" Some abortion critiques object to all abortions regardless of whether you are terminating a 5–8 week of pregnancy shrimp or a human as described in these articles. It is very clear from this report if you are performing a late termination, after 32 weeks that fetus is really a human, terminating is probably human murder. Based on the science presented in this report, the fetus is not really human until at least 32 weeks, terminating a fetus before 32 weeks may be acceptable but it is not acceptable after that point.

What happens to the human fetus? The human fetus grows, develops organs seemingly advancing as fetal age advances. The human fetus also strangely passes through a form of evolution starting as a single cell species, then becoming a multicell species, then a simple eukaryote, then a shrimp-like species, and then on to a mammalian species. With each advance, the brain and nervous system develops and species becomes more and more advanced ending up at 32 weeks, resembling a human, able to think and remember (see Chapter 23).

FIGURE 24.1   Japanese fetal graveyard tomb.

How should a pregnant woman look at her human fetus? Some mothers give 6–8 week fetuses a human name and look at the fetus like a human child. When they die from biochemical pregnancy or miscarriage, some women consider it like a child's death. The only issue is that it not. In Japan, women bury miscarried babies in special cemeteries. Shown in Fig. 24.1 is a miscarried fetus's tomb site. If a child is a biochemical pregnancy and is aborted at six weeks of gestation, it is just a tiny ovular bunch of cells, probably not including any brain cells. It is probably reminiscent of nothing more than bacteria viewable under a microscope. How can this be looked at as a human? If a child is miscarried at 8–10 weeks of pregnancy, it is like a shrimp, with a miniscule 0.00026% of body weight brain, smaller than a shrimp brain (Fig. 24.1). How can this fetus be loved and looked at as a human?

This is all hard to comprehend and hard to understand. Pregnant women need to be taught about the strange growth and evolution that goes on during pregnancy. Yes, the fetus has human genes and yes, the fetus originated from humans, but they really are not biologically akin to humans with a powerful brain.

Index

'*Note*: Page numbers followed by "f" indicate figures and "t" indicate tables.'

A

A-ATPases. *See* Archaea-ATPases
    (A-ATPases)
Abortion, 175
Acetate-based reaction, 12
Acetyl coenzyme A (Acetyl CoA), 46, 67
Acrosome reaction, 144–145, 145f
Adenine, 55
Adenine–thymidine bases, 52
Adenine–uracil bases, 52
Adenosine diphosphate (ADP), 65, 66f, 79
Adenosine monophosphate (AMP), 66f, 68
Adenosine triphosphate (ATP), 12, 30, 37, 65,
    66f, 79
  ATPase, 79
  cytoplasmic membrane electron transport,
    75–76
  energetics
    evolution, 76
    and life, 77
  energy sources, 73f
    of life, 66f
  in eukaryotes, 65
  F-ATPases, 80, 82
  and heart, 72–74
  mitochondria, 81
  oxidative phosphorylation, 66–68, 67f
    feeding pathways to, 68–71
  oxygen generation, 74–75
  photosynthesis, 74–75
  synthesis in bacteria, 75–76, 76f
  use of energy, 71–72
ADP. *See* Adenosine diphosphate (ADP)
Advanced brain animals, 103–104
Alexander Oparin hypothesis, 25
Alveolates, 98
Amino acids, 17, 25, 70
  acidic, 54
  early peptides, 27
  Miller–Urey experiments, 25, 26f
  polymerization, 26
  polymerization, 30
  use of amino acid Alanine, 71f

Aminoacyl-tRNA synthetases, 61
Ammonia, 26
Amnion cells, 147
Amoebae, 96
Amoebozoa, 98
AMP. *See* Adenosine monophosphate (AMP)
Animal evolution, 101. *See also* Human
    evolution
Animal(s), 119
  advanced brain, 103–104
  cells, 97–98
  small brain, 101–102
Archaea, 96
Archaea-ATPases (A-ATPases), 80
Archean eon, 8–9, 85, 86t–92t
Ardipithecus ramidus (*A. ramidus*), 105–106
Asp–Ala–Lys–Val–Gly–Asp–Gly–Asp
    peptide, 27
Asp–Gly–Asp subgroup peptide, 27
Asp–Gly–Asp–Ala–Asp peptide, 27
Asp–Tyr–Asp–Gly–Asp peptide, 27
Asteroids, 6–7
ATP. *See* Adenosine triphosphate (ATP)
ATP synthase (ATPase), 79
Australopithecus afarensis (*A. afarensis*),
    105–106, 114
Autocrine hyperglycosylated hCG, 163

B

Bacteria, 96
  ATP synthesis in, 75–76, 76f
  bacterial membrane system, 76
  cytoplasmic membrane electron transport,
    75–76
Big Bang Theory, 3–4
Billion years ago (BYA), 4, 11
Biochemical pregnancies, 158
Biochemistry of early life
  atmosphere on Earth, 29
  Calvin cycle and glycolysis, 33f
  carbon monoxide–Methane/ATP energy
    system, 30, 31f

Biochemistry of early life (*Continued*)
  early biochemical molecules, 29
  Simple Life-1, 30
  Simple Life-2, 30
  Simple Life-3, 32
Bipedalism, 113–115
Blastocyst formation, 147–148
  embryo growth, 148f
Blastocyst hatches, 157
BYA. *See* Billion years ago (BYA)

C
Calcium permeability, 144
Calmodulin-binding proteins, 144
Calvin cycle, 32, 74
  for converting CO_2 to glucose, 33f
Carbon monoxide–acetate ATP-generating
      pathway, 37
Carbon monoxide–methane/ATP energy
      system, 30, 31f
Cell wall, 41
Cenzoic Era, 85, 86t–92t
CG. *See* Chorionic gonadotropin (CG)
CG–SO₄. *See* Sulfated form of chorionic
      gonadotropin (CG–SO₄)
Chemical life, 45, 93. *See also* Eukaryotic life;
      Prokaryotic life
  evolution, 94f
Chiasmata. *See* Chromatids
Chimpanzee, 103
Chloroplasts, 74, 97
Chorion cells, 147
Chorionic gonadotropin (CG), 108–112
Chromatids, 135–136
Chromosomes, 62
Citric Acid Cycle, 66–69
  use of glucose and glycolysis, 69f
Civilizations, 124
Clays, 16
Cloning, 135
  of spermatogonia, 136f
Coagulation, 6–7
Competitive generation of follicle and oocyte.
      See Oogenesis
Conjoint twins, 148–149, 149f
Copying process, 57–58
Corpus luteum, 127, 132–133
Cowper's gland, 140
Cyanobacteria, 13, 29, 32
  earliest life-form, 15f–16f
Cytodifferentiation of sperm cells,
      136–139, 138f

Cytoplasm, 136–139
Cytoplasmic membrane electron transport,
      75–76
Cytosine, 55
Cytosine–guanine bases, 52
Cytoskeleton, 94
Cytotrophoblast
  cells, 157
  columns, 163, 164f
    extension, 164f
  villi, 163

D
Dark energy, 3–4
  and expanding universe, 5f
Deoxycytidine monophosphate
      (dCMP), 51
Deoxyribonucleic acid (DNA), 55. *See also*
      Ribonucleic acid (RNA)
  biological information, 55–57
  chromosomes, 62
  DNA-based life, 20
    hexanucleotide comparison, 21f
    RNA *vs.*, 20–21
  double helix structure, 57
  double stranded base-paired DNA, 57f
  evolution, 51
    DNA genetic code, 53t
    DNA synthesis, 51–52
    enzymes, 51
    and RNA, 52
  exons, 58
  genes, 62
  genetic code, 57–58
  nucleotide structure, 56f
  replication, 58–59, 58f
  and RNA nucleotides, 55
  transcription, 59–60, 60f
Deoxyribonucleotides (dNTPs), 51
Deoxythymidine triphosphate (dTTP), 51
Deoxyuridine monophosphate (dUMP), 51
Deoxyuridine triphosphate (dUTP), 51
Di-zygotic twins, 149–150
Dictionary. com, 45
Discircristates, 98
DNA. *See* Deoxyribonucleic acid (DNA)
dNTPs. *See* Deoxyribonucleotides (dNTPs)
Dogs, 115
  memory, 103–104
Domains, 96
Dominant follicle, 129–131
  formation of stigma and thinning, 132f

dTTP. *See* Deoxythymidine triphosphate
    (dTTP)
dUMP. *See* Deoxyuridine monophosphate
    (dUMP)
dUTP. *See* Deoxyuridine triphosphate (dUTP)

E

E-ATPases. *See* Extracellular-ATPases
    (E-ATPases)
Early life, biochemistry of
    atmosphere on Earth, 29
    Calvin cycle and glycolysis, 33f
    carbon monoxide–Methane/ATP energy
        system, 30, 31f
    early biochemical molecules, 29
    Simple Life-1, 30
    Simple Life-2, 30
    Simple Life-3, 32
Early life-form, 14, 15f–16f
Ectopic pregnancies, 158
EGF. *See* Epidermal growth factor (EGF)
Einstein's theory, 4
Elephants, 119
Elongation factors, 61
Emerging egg. *See* Ovarian follicle
Endosymbiosis, 96, 97f
Energetics
    evolution, 76
    and life, 77
Energy, 65
Enzyme activity. *See* F1 ATPase complex
Epidermal growth factor (EGF), 147
Epididymis, 139
Epitheliochorial placentation, 101–102,
        108, 108f
    brain size, 109t
Eukaryota, 96
Eukaryote(s), 37–38, 65, 68, 98
    evolution, 99f
        from prokaryotes, 96–97
    plants, 67
Eukaryotic cells, 37–42
    eukaryotic and prokaryotic cells comparison,
        40f
Eukaryotic life, 45, 97–98. *See also* Chemical
        life; Prokaryotic life
Evolution of humans. *See* Human evolution
Evolution timeline, 85
    Earth's creation, 86t–92t
    human evolutionary line, 92
    massive tree stretches, 85
Exons, 58

Expulsion of oocyte. *See* Ovulation
Extracellular-ATPases (E-ATPases), 80
Extravillous cytotrophoblast cells,
        165–166

F

F-ATPases. *See* Phosphorylation
        Factor-ATPases (F-ATPases)
F0 ATPase complex, 81
F1 ATPase complex, 81
FADH$_2$. *See* Flavin adenine dinucleotide
        hydrogen (FADH$_2$)
Fatty acids, 68–69
Feeding pathways to oxidative
        phosphorylation, 68–71
    amino acid Alanine use, 71f
    fatty acids use, 69f
    glucose and glycolysis use, 69f
    glycogen use, 70f
Fertilization, 146–147, 146f
Fertilized human zygote, 173
Fetal cells. *See* Amnion cells
Fetal life, 163
Fetus. *See also* Pregnancy
    cerebral cortex activities, 170
    development during pregnancy, 167, 168f
    fetal structure, 167–168
    fetal weight, 170t
    genes, 168
Five prime (5′), 57
Flagella, 139
Flagellum, 42
Flavin adenine dinucleotide hydrogen
        (FADH$_2$), 67
Follicle stimulating hormone
        (FSH), 127–128, 139
FOXP2 gene, 122
Fructose, 140
FSH. *See* Follicle stimulating hormone (FSH)
Fungal cells, 98

G

GAH-2. *See* Gonadotropin ancestral
        hormone-2 (GAH-2)
Galveston, 173–174
Garbage bags, 12
Genes, 62
Genetic code evolution, 52
    acidic amino acids, 54
    DNA genetic code, 53t
    Human Genome Project, 52–53

Germ cells of immature testis, 135
Gestation, 167–170
Gesture, 122–123
  gestural language, 123
  gestural theory, 123
Gibbons, 120
Glucose, 68–70
Glycogen, 69–70
  as source of energy, 70f
Glycol nucleic acid, 19–20
GnRH. *See* Gonadotropin releasing-hormone
    (GnRH)
Golgi apparatus, 38–40
Gonadotropin ancestral hormone-2 (GAH-2), 110
Gonadotropin releasing-hormone (GnRH),
    110, 130
Granulosa cells, 127
Grex, 42
Guanine, 55

H

"*H. sapiens praesto*", 113
H⁺ ions, 76
H⁺-transporting ATPases. *See* Phosphorylation
    Factor-ATPases (F-ATPases)
Hadean eon, 8, 85, 86t–92t
Haploid, 147
hCG. *See* Human chorionic gonadotrophin
    (hCG)
hCG-1. *See* Human chorionic gonadotropin
    (hCG)
hCG-2. *See* Hyperglycosylated hCG
hCG-3. *See* Pituitary sulfated hCG
hCG-4, 155
Helicase breaks, 58–59
Hemochorial placentation, 103, 110–111, 111f,
    154f, 163, 165–166, 165f
  brain size of humans with, 109t
  cytotrophoblast columns, 163, 164f
    extension, 164f
  cytotrophoblast villi, 163
  hCG, 163, 164f
  villous structure formation, 165f
Hepta-zygotic twins, 149–150
Hexa-zygotic twins, 149–150
Hominids, 105
  brain size, 109t
  evolutionary heritage of humans, 106f
Homo erectus (*H. erectus*), 106
Homo habilis (*H. habilis*), 106, 112
Homo heidelbergensis (*H. heidelbergensis*),
    106–107, 113, 123

Homo neanderthalensis (*H. neanderthalensis*),
    107, 123
Homo sapiens (*H. sapiens*),
    107, 122
  development, 121f
Human brain, 107–113, 167
  development, 167–170
    during pregnancy, 169f
  language function, 121–122
  size, 168t
Human chorionic gonadotrophin (hCG), 130,
    148, 151, 154, 157, 163, 164f, 165,
    174–175
  multiple semiindependent variants, 152t
Human evolution, 105. *See also* Animal
    evolution
  A. afarensis, 105–106
  A. ramidus, 105–106
  back to hominids and
    primates, 106f
  bipedalism, 113–115
  H. erectus, 106
  H. habilis, 106
  H. heidelbergensis, 106–107
  H. neanderthalensis, 107
  H. sapiens, 107
  human brain, 107–113
  O. tugenensis, 105
  S. tchadensis, 105
Human female oogenesis. *See* Oogenesis
Human Genome Project, 52–53
Human male spermatogenesis. *See also*
    Oogenesis
  meiotic division, 135–140
    cytodifferentiation and packaging of
      sperm cells, 136–139, 138f
    endocrine control of spermatogenesis, 139
    male reproductive anatomy, 140f
    maturation of spermatozoa, 139
    seminal fluids, 140
    mitotic proliferation, 135
Human(s)
  brain size, 109t
  conception of life, 47
  development
    human pseudoevolution, 123–124
    language, 119–123
  fetus, 175
  life, 173, 175
Hyperglycosylated CG, 110
Hyperglycosylated hCG, 151–153, 157, 163,
    164f, 165
Hypophysectomy, 139

I

IGF-1. *See* Insulin-like growth
    factor-1 (IGF-1)
IGF-BP. *See* Insulin growth factor binding
    protein (IGF-BP)
Implantation, 157
    approach of uterine epithelium, 158f
    burial deep into uterus, 160f
    invasion process, 159f
Inhibin, 128–129, 129f
Inner cell mass. *See* Amnion cells
Insulin growth factor binding protein
    (IGF-BP), 127–128
Insulin-like growth factor-1
    (IGF-1), 127
Insulin-like growth factor-2 (IGF-2), 127
Intron, 58

J

Japanese fetal graveyard tomb, 176f

K

Lactate, 73–74
Language, 102–103, 119–123
LH. *See* Luteinizing hormone (LH)
Life, 11, 45
    chemical life, 45
    Dictionary.com, 45
    energetics and, 77
    human conception of life, 47
    hypothetical stages, 46f
    ingredient in, 25
    Merriam-Webster.com, 45
    Miller–Urey experiments, 25, 26f
    NASA, 46–47
    peptides on primitive Earth, 26
    RNA, 47
Life-forms, 11
    chemical reaction, 12
    early life-form, 14, 15f–16f
    hypothetical stages, 46f
    ribonucleotide, 17
    "RNA world", 12
    stromatolites, 13
    volcanic environments, 13, 15
Living species, 96
Luteinizing hormone (LH), 110, 129–130,
    139, 153
    on follicular growth, 129f
Luteogenesis, 127, 131–133, 133f
Lysosome, 40–41

M

Mammal evolution, 101, 102f. *See also* Human
    evolution
Mammals, 101, 108
    friendly face, 103f
Meiosis, 130, 136
    stages, 131f, 137f
Meiotic division, 135–140
    cytodifferentiation and packaging of sperm
        cells, 136–139, 138f
    endocrine control of spermatogenesis, 139
    male reproductive anatomy, 140f
    maturation of spermatozoa, 139
    seminal fluids, 140
Memory of dogs, 103–104
Menstrual cycle components, 127
Merriam-Webster. com, 45
messenger RNA (mRNA), 57–58
    splicing, 60, 61f
    translation, 61, 62f
Metabolic remodeling, 74
Methanosarcina acetivorans (*M. acetivorans*), 12
Mezozoic Era, 85, 86t–92t
Microtubules, 41
Middle Paleolithic period, 123
Milky Way, 5–6
Miller–Urey experiments, 25, 26f
Million years ago (MYA), 85
Miscarriages, 158
Mitochondria, 41, 81, 96–97
    mitochondrial matrix, 81
Mitochondrion, 66–67
    oxidative phosphorylation in, 67f
Mitotic division. *See* Cloning
Mitotic proliferation, 135
Monozygotic identical twins, 148–149, 149f
Moon, 7
Morphemes, 120
Morula formation, 147–148
    embryo growth, 148f
mRNA. *See* messenger RNA (mRNA)
Multicellular organisms, 42
Multiple hCG molecules
    hCG, 151, 154
        hyperglycosylated, 151–153
        multiple semiindependent variants of, 152t
        pituitary sulfated, 153–154
    hCG-4 and hCG-4 free β, 155
    hemochorial placentation, 154f
    independent hCG molecules, 151–155
    N-linked oligosaccharides, 153f
    O-linked oligosaccharides, 153f
MYA. *See* Million years ago (MYA)

N

N-linked oligosaccharides, 153f
NADH. *See* Nicotinamide adenine dinucleotide
hydrogen (NADH)
Neanderthals, 123
Neolithic Revolution, 124
Neurons, 167–168, 170t
Nicotinamide adenine dinucleotide hydrogen
(NADH), 67
Nonoxidative pathways, 71–72
Nuclear fusion, 5
Nucleoid, 41
Nucleolus, 37–38
Nucleoside diphosphate kinase, 68
Nucleosomes, 57
Nucleotides, 22, 46–47
Nucleus, 37–38

O

O-linked oligosaccharides, 153f
OH⁻ ions, 76
Oocytes, 127, 130, 147, 149–150
Oogenesis, 127–130. *See also* Human male
spermatogenesis
luteogenesis, 131–133, 133f
ovarian follicle structure, 128f
ovulation, 130–131, 132f
primordial follicles, 128f
Organelles, 41–42
Origin of life, 11–12, 15
Orrorin tugenensis (*O. tugenensis*), 105
Ovarian follicle, 127, 128f
Ovulation, 127, 130–131, 132f
Ovum propulsion, 145–146
OxfordDictionaries. com, 46
Oxidative energy, 72
Oxidative pathway, 71–72
Oxidative phosphorylation, 66–68, 67f
feeding pathways to, 68–71
in mitochondria, 67f
Oxygen
atmosphere, 37
generation, 74–75

P

P-ATPases, 80
Packaging of sperm cells, 136–139, 138f
Paleozoic Era, 85, 86t–92t
Palmitic acid, 68–69
Panspermia theory, 13–14, 25–26
Penta-zygotic twins, 149–150

Peptides
early peptides, 27
nucleic acid, 19–20
on primitive Earth, 26
Phosphorylation Factor-ATPases (F-ATPases),
80, 82
Photosynthesis, 37, 41, 74–75, 75f
photosynthetic membrane, 41
Pilus, 41
Pituitary gland removal. *See* Hypophysectomy
Pituitary sulfated hCG, 153–154
Placental cells. *See* Chorion cells
Placental function, 163
Planet Earth, 3, 5–6, 29
asteroid merging with other asteroids, 7f
atmospheres on, 9f
birth, 6–8
composition, 10f
evolutionary history, 38t, 39f
energy pathways, 39f
eukaryotic cells, 37–42
multicellular organisms, 42
oxygen atmosphere, 37
prokaryotic cells, 37–42
in expanding universe, 3–4
geology and chemistry, 8–9
in Solar system, 4–6, 6f
Plant cells, 98
Plasmid, 41–42
Plasticity of brain organization, 122
Polymerization, 26
Porous membrane, 145
Pregnancy. *See also* Fetus
brain weight during, 169
failure, 158–159, 160f
hCG, 166
hyperglycosylated hCG, 166
testing, 173
Primary oocytes. *See* Primordial follicles
Primates, 105
brain growth genes, 108–109
brain size, 109t
evolutionary heritage of humans, 106f
lower simian, 110–111
Primitive Sun. *See* Protostar
Primordial follicles, 127, 128f
Primordial life, 8
Primordial soup, 11–12
Prokaryotes, 37, 93–94
eukaryote evolution from, 96–97
Prokaryotic cells, 37–42
eukaryotic and prokaryotic cells
comparison, 40f

Prokaryotic life, 45, 93–96. *See also* Chemical life; Eukaryotic life
 evolution, 95f
Proteins, 70
Proterozoic Eon, 85, 86t–92t
Protocells, 30
Proton channel. *See* F0 ATPase complex
Proton motive force, 76
Protostar, 5

R

Radioactive half-life dating, 8
Red cells, 147
Replication
 DNA, 58–59, 58f
 fork, 59
Ribonucleic acid (RNA), 19, 30, 47, 55. *See also* Deoxyribonucleic acid (DNA)
 DNA *vs.*, 20–21
 hexanucleotide comparison, 21f
 hypothesis, 19
 nucleotide structure, 56f
 nucleotides, 22
 riboswitches, 21–22
 short RNA molecules, 20
 transfer molecule, 22
 world, 12, 19
  hypothesis, 22
Ribonucleotides, 17, 25
Ribosome. *See* Ribozymes
Riboswitches, 21–22
Ribozymes, 19
 DNA-based life, 20
 riboswitches, 21–22
 RNA world hypothesis, 22
 structure, 19
RNA. *See* Ribonucleic acid (RNA)

S

Sahelanthropus tchadensis (*S. tchadensis*), 105
Secondary follicle, 127
 formation, 127
Semen, 140, 143
 components, 141f
Seminal fluids, 140
Sense strand, 60
Septa, 98
Shorter pili, 41
Simple Life-1, 30
Simple Life-2, 30
Simple Life-3, 32

Small brain animals, 101–102
Smooth endoplasmic reticulum, 38–40
Solar Nebular Disk Model (SNDM), 4
Solar system, 4–6, 6f
Sorbitol, 140
Sounds, 120
Species, 119
Sperm
 activation, 143
 capacitation, 143
  spermatozoa acrosome reaction, 145
  spermatozoa capacitation, 143–144
 sperm cells, cytodifferentiation and packaging of, 136–139, 138f
Spermatids
 cytodifferentiation, 138f
 early round, 136–139
Spermatogenesis, 135, 139
 endocrine control of, 139
Spermatogonia
 cloning of, 136f
 syncytium, 136f
Spermatozoa
 acrosome reaction, 145
 capacitation, 143–144
  spermatozoon and capacitation, 144f
 maturation of, 139
 motility, 144
Spermatozoon, 138f, 144f, 146–147
Spiral uterine arteries, 163
Spontaneous abortions. *See* Miscarriages
Stromatolites, 13, 29
Sulfated form of chorionic gonadotropin (CG–SO$_4$), 130
Symbiosis, 96
Syncytiotrophoblast
 cells, 157
 tissue, 163, 165–166
Syntax, 120

T

T-DNA. *See* Thymidine-DNA (T-DNA)
Testis, 136–139, 137f
 germ cells of immature, 135
Tetra-zygotic twins, 149–150
TGF α. *See* Transforming growth factors-α (TGF α)
Theca cells, 127
Three prime (3′), 57
Threose nucleic acid, 19–20
Thylakoid membrane. *See* Photosynthetic membrane

Thylakoids, 74
  in chloroplast in plant, 74f
Thymidine-DNA (T-DNA), 51
Transcription, 57–60, 60f
  factors, 60
Transfer RNA (tRNA), 57–58, 61
Transforming growth factors-α (TGF α), 147
Transforming growth factors-β (TGF β), 110,
  147
  receptor, 151–153, 155, 157
Translation, 57–58
  of mRNA, 61, 62f
Tri-zygotic twins, 149–150
Triple dominant follicles, 127
tRNA. *See* Transfer RNA (tRNA)
Tubal pregnancies. *See* Ectopic pregnancies
Twinning, 148–150
  monozygotic identical twins and conjoint
    twins, 149f

U
U-DNA. *See* Uracil, DNA containing (U-DNA)
Universe, 3–4
  dark energy and, 5f
Unovulated primary follicles, 127

Upper Paleolithic Age, 123–124
Uracil, 55
Uracil, DNA containing (U-DNA), 51

V
Vacuole-ATPases (V-ATPases), 80
Vervet monkeys, 119
Vocal language, 123
Volcanic environments, 13, 15
Volcanic sites, 29–30

W
Walking, 114
Wikipedia. com, 46

Y
Yellow cells. *See* Chorion cells

Z
Zona pellucida (ZP3), 145, 148–149
Zonary placentation, 103
  mammals brain size, 109t
Zygote, 147